SERIOUS
GAMES

Clark C. Abt

SERIOUS
GAMES

NEW YORK

The Viking Press

Viking Compass Edition
Issued in 1971 by The Viking Press, Inc.
625 Madison Avenue, New York, N.Y. 10022

Distributed in Canada by
The Macmillan Company of Canada Limited

SBN 670-63490-5 (hardbound)
670-00313-1 (paperback)
Library of Congress catalog card number: 79-83234

Printed in U.S.A.

Fourth printing February 1974

Acknowledgments

Advanced Research Projects Agency: for "Politica." Central Michigan University, Central Michigan Education Research Council: for "Sepex." Harvard University: for "Machinist Career Simulation"; this work was supported by a subcontract between Harvard University and Abt Associates; this subcontract was financed by Grant #PEG-1-6-061819-2240, a contract between Harvard University and the U.S. Office of Education. D. C. Heath and Company: for "Equation Match," "Equation Rummy," "Equals," and "Multiplication and Division." Joint War Games Agency: for "Temper." Price Waterhouse & Co.: for "Merger" (Capital Budgeting Simulation). Science Research Associates, Inc.: for "Colony" from *American History Games.* © 1970, Abt Associates, Inc.; all rights reserved; reprinted by permission of Science Research Associates, Inc. Supermarket Institute: for "Superb" and "Supra." Wellesley School System, Curriculum Development Center: for "Neighborhood."

To my colleagues
at Abt Associates

AUTHOR'S NOTE

"The Sun Game," "Weather," "Color," and "Shape" were designed by Martin S. Gordon.

"Empire" was designed by Robert Branfon and the author.

"Bushman Exploring and Gathering" was designed by Raymond E. Glazier, Jr.

"Caribou Hunting Game" and "Seal Hunting Game" were designed by Holly Kinley.

"Transportation" was designed by Stephen Bornstein.

"Manchester" was designed by Martha Rosen and John Blaxall.

"Machinist" was designed by Elinor Gollay.

"Corridor" was designed by Martin S. Gordon and James C. Hodder.

"Politica" was designed by Martin S. Gordon and Daniel DelSolar.

"Superb" was designed by James C. Hodder and W. Randolph Folks.

"Bankloan" was used in a banking seminar to demonstrate the use of games in bank training programs; it was designed by Richard Rosen, the author, and Richard Scott.

"Supra" was designed by Ted I and W. Randolph Folks.

"Colony" was designed by Alice K. Gordon.

CONTENTS

Preface   xiii

I. *The Reunion of Action and Thought*   3

The uses of abstraction. What are games? Games and
reality. Games among children; among adults. Warfare. The serious game in industry, education, and
government. Games and knowledge. Games as motivation and communication. Role-playing.

II. *Improving Education with Games*   15

The gap between education and life. Student motivation. Simulating complex processes. The mathematical
theory of games. Conflict strategies. Intuitive problem-
solving. How students learn from each other. Developing social behavior. Games as a measure of untested
abilities. Involving the student in game design and

preparation. Selecting topics for classroom games.
The teacher as observer, analyst, critic. Post-mortem
game analysis. Timing of classroom games. Scoring.

*III. Educational Games for the Physical and
Social Sciences*   35

Differences between games for the social studies
and for the sciences. Games for elementary science:
"Weather," "Color," "Shape." Mathematical games.
The importance of decision-making in social-studies
games. Elements common to social studies and to
games. War games. "Grand Strategy." "Empire."
Moral questions raised by games and discussed by
the students. Various "Hunting" games simulating
primitive cultures.

IV. Game Learning and Disadvantaged Groups   61

Problems of ghetto children in the educational world.
How game dynamics can reduce cultural barriers and
student apathy. Tailoring games to particular levels
of intelligence and aptitude. Game-playing as an immediate reward. The supportive role of the teacher.
"Raid," a classroom game in which students play the
roles of police, racketeers, and the city population.
"Manchester," a game about the Industrial Revolution in England. "Neighborhood," a game illustrating
urban growth.

V. Games for Occupational Choice and Training   79

Guidelines for developing games simulating specific
occupations to help students choose careers. "Machinist."

*VI. Games for Planning and Problem-Solving
in Government and Industry*   89

Games as a mode of experimenting with different
strategies in solving a problem. The importance of

role-playing. Three games for government: "Corridor," exploring alternative plans for Northeast Corridor transportation; "Politica," portraying a prerevolutionary crisis in Latin America; and "Simpolis," dealing with urban problems. Three games for industry: "Merger," concerning industrial mergers; "Superb," in which players assume the roles of supermarket executives; and "Supra," another supermarket game designed to train players in scientific purchasing.

VII. *How to Think with Games by Designing Them*   103

How games can be developed to clarify personal or domestic problems. Example: Should the family vacation in the mountains or at the sea shore? Identifying the actors' objectives. The importance of empathy in resolving conflicts.

VIII. *How to Evaluate the Cost-Effectiveness of Games*   110

Criteria for assessing the value of games as compared to other instructional and research methods. Active involvement of players. Realism of situation. Clarity of consequences. "Playability" in terms of materials, space, and time required.

IX. *The Future of Serious Games*   119

The uses of games in a technological society. Increased demands on the schools. Making education more effective in the ghettos and rural slums. "Laboratory schools" of the future. The teacher shortage. Education needed for adult dropouts. Games for community-action planning. Training games for police academies. Involving the poor and the disadvantaged in society. A new language of action.

Appendices:

A. Elementary Mathematics Games   135

B. A Game for Planning an Educational System   147

C. "SEPEX": A School Electronics Planning Exercise   157

D. "Colony": A Secondary-School Game   164

PREFACE

The serious games in this book evolved out of my long dual fascination with scientific problem-solving on the one hand and dramatic human conflicts on the other—with science and the humanities and how to combine them. I had a difficult time integrating these two life interests, both in college where only marginal alternatives were offered to the study of mostly engineering or mostly humanities, and later in jobs of systems engineering during the day and creative writing and editing at night. My first experience combining these two apparently irreconcilable interests was in the Air Force, using operations analysis and war gaming for mission planning. This work combined the writing of dramatic scenarios with mathematical analysis and the interplay of large groups.

In the late 1950s engineers and planners concerned with large systems problems such as air defense or mass transportation turned to computer simulation to gain insights into these processes. The overlapping techniques of operations analysis, operations research, and systems analysis were used to develop mathematical models of these vast and complex processes. This was not exactly science, and it is not yet a science today, because so many uncertain factors must be oversimplified and simulated not by exactitudes but by probabilities—that is, with dramatic suspense. And it does require creative effort not unlike a playwright's to design problem situations with dramatic scenarios that best reveal solutions.

My colleagues and I designed computer simulations of air battles, space missions, missile exchanges, disarmament inspection systems, and international political-economic competitions in the Advanced Studies Department at the Raytheon Company Missile Systems Division, which I managed at the time. By the early 1960s our systems analyses and mathematical simulations of antiballistic-missile (ABM) defense systems convinced me that no satisfactory defense was feasible. This led directly to my working on arms control and disarmament, since advanced-weapons technology seemed to me increasingly unproductive of security.

In the arms-control work, problems of international conflict required political and economic analysis much more than they did engineering skill, so I returned to M.I.T. to learn as much as I could about the social sciences and what they could contribute to policy analysis directed toward world peace. I studied arms control problems under Professors Kissinger and Schelling at Harvard, and political science under Professors Pool, Kaufmann, and Wood at M.I.T. My doctoral dissertation was an attempt to identify effective means of terminating wars.

In the course of my political-science studies at M.I.T., the domestic social problems of education, urban and rural economic development, and technology planning and forecasting captured my interest. My inability to pursue these problems in a company oriented toward military hardware led me to found Abt Associates Inc. in 1965. My idea was to work on domestic, nonmilitary problems combining the computer-simulation and war-gaming techniques I had learned in the aerospace industry with the methods of the social sciences.

At Abt Associates I gathered together a group of young systems engineers, mathematicians, economists, political scientists, sociologists, anthropologists, historians, and psychologists who wanted to work on social problems in a less institutional atmosphere than could be found in government, the universities, or big business. Our work in educational-curriculum development, school-system planning, economic development, industrial management, and technological planning and forecasting required a technique for integrating complex interactions quickly and clearly. Simulation modeling, in game or computer model form, proved very useful for analyzing social problems of the most diverse kinds. The simulation games also turned out to be effective motivational factors in our work with managers, students, teachers, and disadvantaged groups. Although most of our one hundred and fifty man-years of applied social research has not been directly devoted to developing simulation games, we use the technique regularly and effectively for problem analysis, education, training, and planning.

I am grateful to Professors Jerome Bruner and Thomas Schelling of Harvard University, James Coleman of Johns Hopkins University, and Elting Morison of Yale University for their early encouragement of my simulation gaming work for educational purposes.

Many of the games described in this book were designed or contributed to by my colleagues at Abt Associates Inc. The creativity and social-problem solving of the following designers of serious games are recognized with particular gratitude: James Barker, John Blaxall, Stephen Bornstein, Peter Crane, Louis Cutrona, Daniel del Solar, Charles Fisher, Raymond Glazier, Elinor Gollay, Alice Kaplan Gordon, Martin Gordon, Grover Gregory, Stephen Guisinger, William Hamilton, James Hodder, Ted I, Joanna Kennedy, Holly Kinley, Emily Leonard, Peter Merrill, Peter Miller, Robert Rea, Martha Rosen, Richard Rosen. Warm thanks are due Mrs. Heidi Gilliam, Miss Mary Gillis,

Miss Billie Renos, and Mrs. Joan Seville for careful and devoted typing of the manuscript and to Mrs. Virginia Cogger and Mrs. Elisabeth Sifton for perceptive editing of a difficult manuscript and for many helpful suggestions. Mr. Alan Williams, editor of The Viking Press, first and always gave me his amiably insightful support.

CLARK C. ABT

Cambridge, Massachusetts
October 1969

SERIOUS
GAMES

I

THE REUNION OF
ACTION AND THOUGHT

As CIVILIZATION evolves toward highly technological societies, the ability to use abstractions becomes more and more necessary for people to function effectively. Only a few dozen generations ago, most men could manage their personal affairs without any knowledge of the abstractions of written words and numbers. Today in Western urban society, it is difficult for an illiterate person to function at even the simplest level, for he cannot read maps or street signs, he cannot add the costs on a grocery list, and he cannot find out through newspapers what is happening around and to him. Only four or five generations ago, simple arithmetic satisfied the calculation needs of even professional men; today even two years of college calculus is often not enough.

The increased emphasis on abstract knowledge is reflected in educational goals, which continue to rise to meet the demands of the society. Frequently, these demands reflect one generation's conception of what a younger generation must know in order to be occupationally successful within a specialized, technological society. These goals are so concerned with adult society, so directed toward a future world which students cannot yet perceive, that the motivation for learning often lags. This is particularly true of many impoverished students who have little contact with educated adult society outside the schools. Often, too, adolescent students fail to see how the abstractions taught them are relevant to the real world.

The world of the child and the young adolescent, like that of primitive man, is a dynamic world, full of emotion, imagination, and physical action. Yet the opportunities for students' expressive action in the schools have yielded to strong pressures from adults for emphasis on abstraction; when students have tried to relate abstract thought to concrete action, adults have frequently felt their own world threatened. Many adults, too, feel frustrated in their need for action and participation in a world which has become increasingly specialized and in which action has come to be the province of specific occupational groups.

Physically inactive thought (mistrusted by Nietzsche) and mentally inactive action (mistrusted by all sensible men) are diseases of civilized man. He often wars without thinking and thinks without reward. Physical action in affluent civilization today is reserved for brutalities, chores, and play. Mental action is practiced chiefly by the physically inactive.

The classic Greek ideal embodied both thought and action, both individuation and participation, *in the same activities.* Today, intense participation in social decision-making is limited to a few individuals appointed, elected, or permitted to represent the larger society. Yet individuals can once again become involved, and thought and action can again be integrated, in games created to simulate these social processes. The zest for life felt at those exhilarating moments of history when men participated in effecting great changes on the models of great ideas can be recaptured by simulations of roles in the form of serious games.

"In dreams begin responsibilities," said the poet, and in games begin realities. Games offer expanded possibilities for action in a mode that, while chiefly mental, includes the felt freedom, intuitive speed, and reactive responses of physical movements.

The word "game" signifies one of those incredibly rich concepts of human activity that have many roots and implications. Even the barest dictionary definition suggests protean significance. "Amusement, diversion." "Fun, sport"—with sport suggesting a physical activity combined with an entertaining mental one. "A procedure for gaining an end," as in playing a waiting game. "A field of gainful activity," like the newspaper game. "A physical or mental competition conducted according to rules with the participants in direct opposition to each other," a general but misleading definition since the participants need not be in direct opposition. "A situation involving opposing interests given specific information and allowed a choice of moves with the object of maximizing their wins and minimizing their losses," a useful, if incomplete, economic definition. As a verb, "to play for a stake." As an adjective, "having a resolute unyielding spirit," presumably in some serious or sporting conflict.

Consider the words appearing in these definitions: amusement, sport, procedure, gain, activity, competition, rules, participants, opposition, information, moves, object, maximizing, minimizing, win, lose, play, spirit. Some of these words describe the formal structure of games: procedure, rules, participants, information, moves, winning and losing. Others suggest the motivation of the participants: amusement, gain, competition, opposition, maximizing, minimizing, play, spirit. Yet all purposeful human activities involve participants, rules and procedures, success and failure. And indeed, the wide use of "game" as a metaphor for many social, economic, political, and military activities shows how much we assume about the formal similarity between games and real-life activities.

My own quite impressionistic view is that a game is a particular way of looking at something, anything. This "way of looking" has two main components, a rational, analytic one and an emotional, creative, dramatic one. The game's analytic component sees in certain aspects of life—family, love, friendship, education, profession, commerce, war, politics, partying—common formal or structural characteristics identical to those of games. These formal characteristics include some or all of the game elements of competing actors or roles or players, objectives or goals that are usually unattainable by all actors at once, resources and powers that the actors use to try to gain their objectives, rules or laws or customs that limit or handicap the actions of the actors and tend to balance them so that the outcome tends to be uncertain, and definite conceptions of winning or losing that are important but mercifully impermanent.

The emotional, creative, dramatic component of the game is made up of a curious combination of optimistic beliefs in the luck of "another chance" and a pessimistic respect for the odds, the chanciness of it all. It is basically an existential view of man's acting, despite uncertainty, to achieve conflicting goals that end up mattering less than the action itself. "It's how you play the game. . . ." Ignorance need be no bar to action. It is, also, pessimistic (and romantic) in its view of life as conflict—with others, with nature, with self—but always unresolved oppositions, uncertainties, overcomings of obstacles. And it offers a kind of spiritual conquest of all evils by incorporating them into stimulating adversary roles that are as necessary to the good as the black is to the red—something the religion game developed long ago.

Games also include a no-nonsense operational ethic that allows no real excuse for losing. It has respect for "the money ball-player" who delivers despite obstacles and knows there is no justification—in game terms—for not doing so. This is an ethic of personal responsibility, in which there are no real excuses except "bad luck," which is a viable excuse only once in a while.

Reduced to its formal essence, a game is an *activity* among two or more independent *decision-makers* seeking to achieve their *objectives* in some *limiting context*. A more conventional definition would say that a game is a contest with rules among adversaries trying to win objectives. The trouble with this definition is that not all games are contests among adversaries—in some games the players cooperate to achieve a common goal against an obstructing force or natural situation that is itself not really a player since it does not have objectives.

Of course, most real-life activities involve independent decision-makers seeking to achieve objectives in some limiting context. The autonomy of human wills and the diversity of human motives result in gamelike forms in all human interactions, and in this sense all human history can be regarded as gamelike in nature.

We can speculate about the origins of games and whether they began in magic, religion, war, or commerce. Whatever the origin, the motive was no doubt each individual's persistent hunger for experience beyond his own. Without the experimentation possible in imaginative life, how could one try out the roles of the admired but inaccessible? How could one win and lose a battle and learn what works and what does not? How could one explore the many branchings in the maze of fate?

The developing man, the primitive, and the child attempt to cope with reality by fashioning models of things they feel to be important, simulating what is believed to be (or, what they would like to believe is) reality in a simplified and scaled-down model they can easily understand. It is a natural step to move from playing with models of objects to playing roles. These roles are really behavioral, "working" psychological models. Here, too, children experiment with reality, attempt to understand it, and "try it on" for size. They play the roles of parents keeping house, cops chasing criminals, firemen putting out fires, and soldiers fighting battles. This basically noncompetitive role-playing may be considered as a primitive form of game in which the structure of alternative decisions has not yet developed, and in which only one line of behavior is followed because of its intrinsic fascination. The exciting uncertainty is that of identity rather than conflict outcome, Who am I? rather than Who will win?

Children tire of playing the same role repeatedly and thus change roles and invent new ones (a process limited only by their knowledge and imagination). When they exhaust the variations of roles, the alternative for adding interest by suspense (uncertainty due to not knowing the outcome) is to add complexity of decision. This occurs naturally when children discover they are not alone in the world and that other, independent, little and big beings are making decisions, some of which frustrate their own.

The complexity of decision-making increases proportionately with the role-player's freedom of decision at every "move." The role-player needs a motivating force to make successive decisions under uncertain conditions, and this is supplied by the competitive decisions of an adversary role-player. Thus, a dynamic is created that generates more surprises, suspense, and dramatic interest than simple repetitive and determined playing of a role. Once they are playing with models or roles that emphasize changes of interactions over time (at the cost of their representational characteristics), children are playing true games.

To grasp the enormous cognitive step children take when they move from dramatic role-playing to interactive games, consider the intellectual difference between "role playing" a knight and playing him interactively with an opposing knight on a chessboard. Or consider a parallel development in the theater: in the Czech exhibit at Expo 1967 the audience of a film periodically voted on the outcome of the preceding dramatic episode and thus "competed" with the film writer and each other to determine the outcome of the sequence. The audience thus participated in the drama on several levels—as audience of a film, as contestant against other viewers and playwright, and as creator of the plot.

The adult activity most clearly analogous to games is warfare. Wars are obviously very costly and not practically subject to experimentation. But they are competitive activities on the largest scale, in which adversary decision-makers contest objectives within the limits of their will and resources. And wars are very complex processes whose outcomes are usually very uncertain (were the outcome certain, few certain losers would fight). In these circumstances no sensible general can afford to plan strategies and tactics, without considering what his adversary might do to counter his plans and what *he* might do to counter the countermeasures. Thus a working model of these reciprocal processes is an essential element in military planning and training.

There is no reason why the learning, analysis, and planning of such processes in the form of games should be limited to military problems. Political and social situations can often also be viewed as games. Every election is a game. International relations are a game. Every personal argument is a game. And almost all business activity is a game. Whether these contests of politics, war, economics, and interpersonal relations are played with resources of power, skill, knowledge, or luck, they always have the common characteristics of reciprocal decisions among independent actors with at least partly conflicting objectives.

Then what is *not* a game? Things are not games. Intrinsically noncompetitive processes, such as production lines, are not games. Predetermined procedures are not games. While all games simulate something from the real world, not all simulations are games. For example, computerized simulations of traffic flow or chemical reactions are not games because their outcomes, while complex to calculate, are predetermined, and there is no winning or losing outcome, only a set of results. It is possible, of course, to make a computer simulation of a game; one such is described later in this book.

Games may be played seriously or casually. We are concerned with *serious games* in the sense that these games have an explicit and carefully thought-out educational purpose and are not intended to be played primarily for amusement. This does not mean that serious games are not, or should not be, entertaining. We reject the somewhat Calvinistic notion that serious and virtuous activities cannot be "fun." If an activity having good edu-

cational results can offer, in addition, immediate emotional satisfaction to the participants, it is an ideal instructional method,
motivating and rewarding learning as well as facilitating it.

The term "serious" is also used in the sense of study, relating
to matters of great interest and importance, raising questions not
easily solved, and having important possible consequences. None
of these aspects of serious games need be associated with their
customarily heavy behavioral baggage of piousness and solemn-
ity. Games may be significant without being solemn, interesting
without being hilarious, earnest and purposeful without being
humorless, and difficult without being frustrating. They may
deal with important behavioral problems, and they may concern
substantive problems in almost all academic and intellectual
fields.

Education, industrial and governmental training, planning, research, analysis, and evaluation are all rich fields for the use of
serious games. In education, games are used by teachers for
classroom instruction in social studies, sciences, and humanities,
and for guidance counseling. They are used by education administrators, such as school superintendents and principals, for the
planning of curricula, assignment of teaching staffs, and allocation of facilities and equipment.

The government uses games to train soldiers, A.I.D. representatives, Foreign Service officers, Peace Corpsmen, and schoolteachers. Planning games are also used to test alternative military strategies, to evaluate regional transportation plans, and to
determine public responses to weather forecasts. In industry,
games are used to train management in complex decision-making, accountants in financial management and merger-acquisition
analysis, bankers and businessmen in loan decisions, and adver-
tisers in media management and sales personnel in basic communications skills.

Complexity and incomplete information characterize many of
the problems of industry and government, yet plans and decisions must be made. To actually test and evaluate various poli-

cies is too expensive in many cases and too risky in others. Games, however, provide a means of identifying and evaluating the consequences of alternative plans and policies.

The physical scientist or engineer experiments in a simulated reality with reduced-scale models of new devices or processes. He builds a replica of a real-world device, such as a model airplane, and tests it in a scaled-down, simulated environment, such as a wind tunnel. He uses a theory about how things are related to each other to define the model relationships, and then experiments with the model to test various solutions in alternative possible environments.

Similarly, problems such as industrial production or traffic flow or chemical processes can often be described in precise mathematical terms, and a mathematical model can be used to test various solutions to various problems. When mathematical models are very complex and involve great amounts of data, computer simulations exercising such models save time and effort. If, instead, a qualitative process is to be analyzed, or one that is still only partly describable in numbers (as are most large social problems), a logical model may be used to refine the understanding of the problem by showing the consequences of various assumptions and solutions. A logical model is often exercised most effectively when used by human players as a kind of game. In such a game, the participants learn the logic of the process they are studying by participating in it and seeing the consequences of their decisions.

The need to experiment inexpensively and creatively is pervasive. Most people experiment with psycho-social situations throughout their lives in ways having most of the elements of games. What we are discussing here are the extended and novel applications of a very traditional developmental mode of human behavior. The oxymoron of *Serious Games* unites the seriousness of thought and problems that require it with the experimental and emotional freedom of active play. Serious games combine the analytic and questioning concentration of the scientific viewpoint with the intuitive freedom and rewards of imaginative, artistic acts.

The abstract representation of real life in game form does not render the game any less capable of teaching "true" knowledge. One does not have to be Shakespeare to understand his plays (which are, after all, monumental literary games), but acting in the plays can yield a more vivid and lasting view of Shakespeare than would a teacher's reading of the plays to a class.

Jean Piaget has said, "Knowledge is not a copy of reality. To know an object, to know an event, is not simply to look at it and make a mental copy, or image, of it. To know an object is to act on it. To know is to modify, to transform the object, and to understand the process of this transformation, and as a consequence to understand the way the object is constructed. An operation is thus the essence of knowledge, it is an internalized action which modifies the object of knowledge." He adds, "Intelligence is born of action," and "Anything is only understood to the extent that it is reinvented." * People in daily life constantly invent and reinvent situations in order to learn from them. Yet too often people fail to recognize that reinventing a situation in which one has been an actor and perhaps reliving or revising decisions made is, in effect, to play a game. People tend to look for an abstract pattern within the situation or to compare situations in order to come to some new abstract conclusion. Yet they often fail to realize that it is an active situation which has led to their new abstract knowledge.

The preoccupation with abstraction which technological society has transferred to its educational systems often destroys or harms an active learning environment. For many traditional reasons, and because of spurious psychological theories and moralistic authoritarianism, a great many elementary and secondary schools still require little more active participation of students than an occasional response to examinations. But a simulation, or

* Quoted in "Jean Piaget: Notes on Learning" by Frank G. Jennings, *Saturday Review*, May 20, 1967.

simplified representation of reality, is always involved in a game, and the game itself becomes an active, dynamic, and changing learning environment.

Perhaps surprisingly, industrial and military training programs, less burdened by learning theories and concerned less with instructor authority than with student performance, have gotten generally better educational results by involving students in actions that reproduce or simulate actual problem-solving situations. This is probably because industry is more "output-oriented," concerned with obtaining specific capabilities at minimum cost, and because there is usually less of an age gap and social gap between teachers and students. Another reason for the apparently higher instructional effectiveness is that the subjects dealt with tend to be concrete, physically demonstrable operations. The Job Corps, for example, has found that industrial training courses in truckdriving, construction, electronics, and food service achieve much higher student interest and performance than the academic subjects do. My impression has been, after interviewing dozens of high-school dropouts at several Job Corps centers, that the active problem-solving activities and the coachlike instructional style of the teachers contributed to educational success.

Games are effective teaching and training devices for students of all ages and in many situations because they are highly motivating, and because they communicate very efficiently the concepts and facts of many subjects. They create dramatic representations of the real problem being studied. The players assume realistic roles, face problems, formulate strategies, make decisions, and get fast feedback on the consequences of their action. Also, with games one can evaluate the students' performances without risking the costs of having errors made in "real-world" tryouts and without some of the distortions inherent in direct examination.

In short, serious games offer us a rich field for a risk-free, active exploration of serious intellectual and social problems. In games man can once again play the exciting and dynamic roles he always enjoyed before society became so compartmentalized. The role-playing that students undertake in games that simulate life is excellent preparation for the real roles they will play in society in later life.

II

IMPROVING EDUCATION
WITH GAMES

In CONTEMPORARY formal education, there appear to be many serious gaps between what is considered worth knowing and what is needed for an effective life. This discrepancy between instructional practice and postschool practicality is attested by the repeated worldly success of mediocre students and the quite modest worldly accomplishments of many outstanding students.

There are even more serious gaps between how knowledge and methods—worthwhile or not—are taught in the schools and how they are most effectively learned in life, indicated in part by the abandonment of free high-school education by over one-quarter of the teen-age population that voluntarily drops out before graduation. America's educational system suffers from motivational, scholarly, intuition-building, social-behavior-training, evaluation, research, planning, and program-development inadequacies.

The motivational inadequacies are probably in most urgent

need of repair. To be motivated is to have a reason for action. The first-grade student, upon entering school for the first time, brings with him a set of values and "reasons for action" established by his parents. He may be positively motivated toward learning, negatively motivated, or simply unmotivated. Though the educational system can never hope to negate the influences of home environment, it can modify these influences in many respects. The highly motivated student can overcome the most unimaginative school curriculum, the most banal school texts, and the most limited facilities if he has the encouragement of his family and his teacher. But in many cases, the educational environment being so restrictive, the teacher feels helpless and becomes unmotivated himself, simply waiting for the end of the school year to find a better school or a better job. In these cases, when the student moves on to higher grades, he is often far less motivated than he once was. But it is obvious that the negatively motivated or unmotivated student suffers most. For if he is not provided with imaginative and exciting "reasons for action" within the school, he will soon find school a waste of time—it interferes with his "reasons for action" outside of school. It is not relevant to his daily life and has not been made relevant by the curriculum, texts, or, most importantly, the teacher.

This problem of motivation, particularly in the elementary grades, becomes more significant when one realizes that the attitudes and skills developed in young children determine their later performance. Most students who do poorly in high school could have been identified early in elementary school.

Students who are not motivated to learn in school are frequently highly motivated in their other activities. Even the youngest children play with a vengeance cops-and-robbers, hide-and-seek, cowboys-and-Indians, and other competitive games. The differences in these two environments, at least in the degree of attention and interest in participating, are partially explained by the great drama in the play environment and the frequent lack of drama in the school environment. This "drama" involves *conflicts* of *uncertain outcome* among actors with whom the child can identify; in playing games, children "become" the characters they represent, and engage vicariously in the conflicts their roles afford. The application of these exciting elements to activities in the school can stimulate the child to learn new intellectual concepts.

Most social, economic, and historical material is full of conflicts of interest. The student's identification with the characters or groups involved in such real-life problems is usually rapid and strong, since students so often deal with uncertainty and conflict in their own lives. And planning, playing, and analyzing educational games motivates the students to study the issues dramatized in the games and the literature dealing with these issues, to synthesize solutions to the problems posed, and to evaluate critically the solutions developed in the process.

I was once given a particularly impressive indication of this motivational effect of simulation games. It happened in the course of demonstrating a game called "Grand Strategy" to some junior-high-school students and teachers in Vancouver, Washington.° The game simulates some events of World War I.

Ten players represented the chiefs of state of ten major nations involved in World War I. Each player's objective was to achieve his nation's political aims at the least military and economic cost. Each player made ten moves simulating strategic decisions taken semiannually from 1914 through 1918. These moves consisted of players making and breaking alliances, deploying armies and navies, and initiating, responding to, and terminating hostilities. The players were seated facing a chalk-drawn map of Europe in 1914. Also posted were the open-alliance relations and the number and location of army divisions and naval fleets of all the powers.

In the morning's game, the war expanded through the entanglement of alliances, much as in the actual history of the prewar period. Just before lunch, the student players were told that they would get another chance to replay afterward. But instead of

° The demonstration was sponsored by the Northwest Regional Educational Laboratories, Portland, Oregon.

going to lunch, most of them went to the school library to study the history of World War I. It is possible that this added study paid off in problem-solving achievement. In the afternoon game, the students were able to keep the war from expanding and to reach a peaceful compromise solution to the international conflict. It might be said that in that one day they were motivated to learn more about how World War I started and ended, and how it might have been settled better, than they would have been taught by any other technique. John Dewey's ideal of the "active learner" was executed successfully.

The superior motivation of students in educational simulation games is widely recognized.* It is obvious that motivation is necessary but not sufficient for learning. But the educational benefits of simulation games other than motivational are only dimly understood. The training that games provide in intuition-building, problem-solving, and social behavior, for example, are of incalculable value.

Most complex processes in technology, economics, and politics include several subprocesses going on simultaneously, or "in parallel" in time. Assembly lines fed by subassembly lines operating simultaneously, simultaneous production and consumption in a market, and simultaneous competition for votes by two or more political candidates are all examples of these processes.

Describing such simultaneous interactions with printed or spoken words is difficult because it requires considerable abstraction and memorization—the learner has to remember that while A was going on, B was starting, and meanwhile C was doing something else, and D was tripped by B. It is like trying to describe a fast play in football with words to someone who, never having played or seen a football game, has limited powers of abstraction, memory, or retrieval of significant data at the right time.

The best way to learn about the parallel processes in a football game is to play it or, second-best, to simulate playing it by

* See Sarane Boocock and E. O. Schild, *Simulation Games in Learning* (Beverly Hills, California: Sage Publications, 1968).

following it through in imagination. The same holds true for learning about other parallel processes in technology, economics, and politics. The inadequate instruction in parallel processes is often never remedied in many liberal-arts students in college, and the students' inability to comprehend and manipulate such abstractions creates an often impenetrable barrier to their making scientific analyses of complex systems problems. It probably also contributes to simplistic thinking in terms of one or two "causes" rather than a "systems" understanding of complex effects.

Another intellectual skill inadequately taught by conventional methods is allocation of resources. The basic idea is that, as with energy and matter in most instances, at any one time within any one system a constant amount of resources are conserved—more in one place means less in another. The "conservative system" of classical physics is also the fixed-resource budget of the economist—more expenditures on one thing mean less on another. The core issue is that of allocating limited resources to maximize one's objectives.

The economic concepts of satisfying minimum objectives * and optimizing achievement of residual objectives by allocations proportional to returns are rarely taught below the college level. But these concepts are implicit in most games in which limited resources must be allocated among competing objectives (such as in successive bids in poker), and explicit in such economic games as "Manchester" and "Empire" (described in Chapters III and IV), games that have been successfully played by schoolchildren.

The mathematical theory of games itself is to some extent a rigorous treatment of this type of problem. ** And the logical

* In the jargon of the mathematical operations researcher, sometimes called "satisficing," perhaps to combine the suggestions of satisfying and sufficing.
** John Von Neuman and Oskar Morgenstern in *The Theory of Games and Economic Behavior* (Princeton, New Jersey: Princeton University Press, 1953) present the basic logic and mathematics of

analysis of decision-making under conditions of incomplete information or uncertainty is a further development of the mathematical game theory exemplified in most games of chance. This is an intellectual realm that is intensely relevant to most of life's decisions, but it is almost never taught below the college level and often is not even included in college liberal-arts curricula. The elementary study of probability and statistics offered in some high schools is not usually sufficient to relate the logical analysis of uncertainty to the decision-making process.

Perhaps the most seminal and pervasive concepts in mathematical game theory are those of "zero-sum" and "non-zero-sum" games, and the "social" or Pareto optimum. A zero-sum game is one in which the sum of the winners' and losers' payoffs (or gains) is always zero. This means that if one player in a two-person game (the mathematics of games explodes in complexity with increasing numbers of players) wins a certain amount, the other player must lose that same amount. If there are three players, and two win, the third player must lose the sum of the winnings of the other two. A non-zero-sum game is one in which the winner's gain is not necessarily at the cost of the loser. Both (or all) players can win, as in peace-keeping, or both can lose, as in nuclear war. Non-zero-sum games are more complex than zero-sum but more like life in that while encompassing the purely competitive aspects they also include the preservation of that game itself which is the "social" and mutual objective of the players. The best strategy in such games is one that maximizes the total wins of all players. This is sometimes called the "Pareto optimum," after the great mathematical economist Wilfredo Pareto, who first expressed this concept.

Conflict strategies are also a common aspect of human civilization that rarely receive attention in the classroom. Mathemati-

both serious (such as duels) and nonserious (such as poker) games. The book, a mathematical and philosophical classic, is not exactly light reading, although much of it is comprehensible to any interested reader.

cal game theory, oligopoly theory of the business firm, and political-military strategic analysis all deal with conflict or competition for incompatible (zero-sum) objectives by adversaries. But competitive processes are also a part of everyday life, and an everyday part of uncommon but crucial moments, such as negotiations among lovers, labor and management, and governments.

It is no accident that the theory that provides a scientific, logical, and quantitative analysis of competitive processes is called "game" theory. Games are the formal equivalent of these competitive processes, stripped of most incidental details. Reducing large-scale competitive processes to simulation games exposes their essential dynamics with a lucidity and drama unequaled by other teaching techniques.*

Intuitive problem-solving is an aspect of education neglected almost everywhere except in multiple-choice tests. It is as if educators considered intuitive problem-solving the moral equivalent of uncontrolled bohemianism. Beats, hippies, and the alienated of all sorts extol intuition with an exaggeration matched only by the disdain many formal educators have for it. It needs to be understood that intuition, while no substitute for knowledge and education, is a valuable application of knowledge: generally, the quality of an intuitive insight is directly related to the amount of knowledge brought to the problem.

A naïve but useful definition of intuitive problem-solving might be that it is the kind of mental activity which is not self-conscious and is therefore difficult to reproduce or explain on demand, but which nevertheless occasionally integrates diverse

* The outstanding work illustrating the applications and limitations of game theory is T. C. Schelling's *Strategy of Conflict* (Cambridge, Massachusetts: Harvard University Press, 1960). In this seminal book, Schelling applies mathematical economics game theory concepts to duels, competitions, and wars. Another lucid treatment—less value-free but useful as an introduction to the subject—is Anatol Rapoport's *Fights, Games, and Debates* (Ann Arbor: University of Michigan Press, 1960).

facts and ideas in a fruitful way not previously apparent to rational, systematic analysis. As such, intuition is a very efficient but somewhat unreliable aid to problem-sensing and problem-solving. To become an effective learning tool, it needs to be stimulated, given expression, tested, and evaluated. Effective intuitions should be rewarded, while incorrect intuitions should be so determined by rational analysis and then discarded. In short, intuition should be developed to complement and advance rational analysis. In their fear of a romantic worship of intuition to the exclusion of rational analysis, educators have often rejected intuition wholesale.

Simulation games stimulate, reward, and judge intuitions according to pragmatic standards rather than doctrinal ones. Enlightening intuitions are rewarded for their superior problem-solving speed over systematic analysis. False intuitions prove to be ineffective in game play. The ideal problem-solving strategy that emerges for most players combines intuition and analysis—analysis used to check intuition, and intuition used to extend analysis beyond familiar limits.

Closely related to the need for instruction in intuition-building is the need for instruction "custom-fitted" to individual student capabilities. Public education by its very nature must be geared to educating the greatest possible number of students. Most students are near the "average" and find their schools not too badly suited to their needs and abilities. But public schools do not usually meet the needs of students far below or above the average. Many students are discouraged by material that is too difficult for them; superior students are bored by material that is too simple. Providing equal opportunity for all students, regardless of their unequal abilities or limitations, is one of the great difficulties of public education.

The one-to-one relationship between a skilled tutor and a student is, of course, ideal, but it is also impractically expensive. The self-directed learning provided by programed instruction also permits the pupil to proceed at his own pace, but it is cogni-

tively and socially limiting: only the predetermined answers are acceptable, and this tends to limit programed instruction to rote learning of simple facts. The excitement and encouragement generated by another human being is missing, as is the intelligent response to a novel solution. True interaction, which produces the most memorable kind of "feedback" in problem-solving, does not occur.

How then can instruction be individualized in an economical but nonlimiting way? Peer learning, or the direct instruction of students by other students, is one way which appears promising. In the Homework Helper program in New York City, for example, student instructors advanced even more than the students they tutored.

Peer instruction can be made still more efficient and intensive by using educational simulation games and small group projects. Students can simultaneously learn different things on different levels in the same game—and probably learn them better than they would from an older teacher. In one game played in a ghetto school, for example, the students ranged from a nearly illiterate eleven-year-old to an advanced high-school senior. Both learned through the game about social change in an industrializing society, although on different levels of sophistication.

Individual instruction is aided by the many decision alternatives that must be confronted by the players in any effective educational game. Even relatively simple simulation games are sufficiently rich in content to provide several different levels of learning simultaneously to students of different abilities. The slow learners will concentrate on the concrete, static elements of the game. The moderately fast learners will develop concepts of cause and effect and attempt to apply them. The most advanced learners will consider the strategic interactions of several parallel causal chains.

Training students in acceptable social behavior is one of the schools' most important tasks, yet it is rarely accomplished realistically in the conventional classroom. Schools and teachers usually have predetermined values expressed in "Dress Codes," "Rules for the Playground," or "Student Handbooks." The values inherent in these rules or codes are imposed from above, and the student must obey or be punished. There is rarely an opportunity for cooperative problem-solving requiring student leadership and negotiation. These skills can be developed, however, through small-group or team activities in simulation games which improve cooperative social skills in concert, rather than in conflict, with cognitive ones.

Grading "on the curve," * which occurs in most schools, is intrinsically competitive in an unproductive way. It reduces the matter of marks to a zero-sum game: every high mark given to one student makes it tougher for other students to also obtain a high mark. Many students seem to realize this instinctively, and thus achievement is inhibited because it is frowned on as "making things tough for the others." A more socially cooperative way of grading (making classroom learning a non-zero-sum game) would be to distribute the students into small teams competing for objective (rather than relative) achievement scores. These student "learning teams" would encourage cooperative problem-solving behavior among team members, while enjoying the motivating powers of team competition. One of the obvious ways the teams could interact would be in games.

Another type of socialization usually omitted from the classroom is that combination of competitive initiative, objective calculation, and courteous restraint which we call "sportsmanship." This may be learned on the playing fields and on the streets, but not all students learn it this way, and none now learn it in association with intellectual effort. The making of classroom learning into a team sport offers the possibilities of applying the best motivational and socializing values of athletics to competitive intel-

---

* "Grading on the curve" means giving a class of students grades on the basis of their relative position to each other on a statistical distribution of all the raw test scores in the class. The top ten per cent may get an A, the bottom ten per cent an F, and so on.

lectual activities. It may also lend a kind of formal legitimization of adolescent values to the learning process itself.*

The evaluation of students is presently done by the teachers' grading, standardized tests, or a combination of the two. But the teachers may be biased, and standardized tests are given only rarely, possibly when a student is not at his best. On the whole, though, the two methods complement each other—except where they are applied to students for whom they are not designed, such as those of a different cultural background from the teachers and testers. Since conventional English-language skills are needed to do well on these tests, we can expect a non-English-speaking foreigner to perform poorly. Likewise, we can expect the "disadvantaged" student speaking a "public" or "nonstandard" English to do badly too. Obviously some culturally less specific instrument is needed for evaluation, and games may offer one such tool.

In the summer of 1965 I met some high-school dropouts who were enrolled in a remedial course at Thompson Academy in Boston. They scored in the 80's on I.Q. tests and performed poorly enough in the classrooms to be regarded as "backward." Yet when several of these teen-agers participated in a game simulating the interactions of city block residents, racketeers, and police, they came suddenly alive and performed not only well, but brilliantly. It was clear to all of us that this performance was an indicator of a kind of problem-solving intelligence that had escaped conventional measurement.

Games could be used by schools to identify specific types of nonverbal abilities—cognitive problem-solving, social negotiating, organizing, and communication skills.

---

* James S. Coleman, in his book *The Adolescent Society* (New York: Free Press, 1961), discusses how the teen-age peer culture tends to inhibit academic achievement, and how athletic contests might be mobilized for academic purposes. Professor Coleman is a leading designer of educational games for teaching the social sciences, as well as a distinguished sociologist.

This use of games to identify the superficially "backward" student who has good unexpressed intelligence is potentially very significant in several ways. Research has shown that many teachers tend to work most closely with students they believe to be "bright" and neglect those they believe to be "dull." * Teachers' expectations often become self-fulfilling prophecies, as the apparently bright students are stimulated further, and the apparently dull are "written off" to mere custodial care. But if games identify otherwise hidden intelligence in a student, he gains another chance for a better rating in his teacher's self-fulfilling success prophecy.

Also, "games testing" has great potential as a means of measuring skills that are extremely important in adult life but essentially untested in formal academic exams. If I were a company president hiring a manager or a general problem-solver, I would much rather observe a candidate's performance in a challenging, multiplayer game than in a conventional test, because the problem-solving needed in the game is much closer to that needed on the job than are the written tests.

Educational planning—of curricula, classroom schedules, instructional methods, teacher recruitment, school facilities and equipment utilization, and allocation of students to schools, etc. —is still largely a cut-and-dried kind of thing. The bold experimentation required is usually too costly in terms of time lost, dollars spent, administrative chaos, and potential damage to students. Experimenting, measuring the results, developing the theory, and verifying or correcting initial hypotheses are limited today largely to analyses of one or a few variables, with *ceterus paribus* assumptions for the rest.** But simulations or games offer an inexpensive and relatively unthreatening means of experimen-

---

* Also known as the Rosenthal effect, after its principal investigator, Professor Richard Rosenthal of Harvard University.

** One notable exception is the U.S. Office of Education report *Equality of Educational Opportunity* (U.S. Government Printing Office, 1966), also known as the Coleman Report after its principal author.

tation. The exercise of "manual" models (human-operated, as opposed to computer-operated) has, in fact, already aided in developing a quantitative, mathematical model, or manipulable theory, predicting student achievement changes, dropout rates, expected average lifetime earnings, and equality of educational opportunity as a result of teacher, materials, and facilities changes.*

In another instance, some forty school superintendents at the Central Michigan Education Research Center played a game the object of which was to acquaint them with the electronic possibilities for improving the efficiency of their schools. Within a few hours they successfully negotiated a plan to link together several diverse school districts with teletypes, educational TV, and a time-shared computer. A conventional conference would have used up days of diffuse talking without ever getting around to these specific solutions.

Another game simulated one crisis-ridden day in an experimental Job Corps center for women. The roles of students, staff, local citizens, and government visitors were simulated by the trainee players. Crises were introduced on a programed basis and were responded to by them. After the simulation was over, they rated each other's performance according to various criteria of administrative efficiency, communicativeness, creativity, crisis management, etc., etc. The game revealed weaknesses of indecisiveness and insensitivity on the part of some candidates who had been very impressive in interviews, as well as the skills of others who did not seem capable in the interviews.

In education, there are no substitutes for highly motivated and creative teachers, relevant and exciting school texts, imaginative and well-planned curricula providing individualized instruction, and effective and well-designed school facilities and plants. But the day has not yet arrived in which all these ideals flourish in all school systems, and until that day does arrive,

* Clark C. Abt, "Design for an Elementary and Secondary Education Cost-Effectiveness Model." (Abt Associates, Inc., Cambridge, Massachusetts, 1967).

games will play an important role in educational life. Games are not a panacea for all the ills of the educational system today, but they do provide fast and effective relief for some of these ills. The challenges posed to educators by the problems involved in planning games force them to deal with new problems and to see old problems from larger or different perspectives. Thus games serve a creative as well as an analytic function in educational planning and programing.

The central idea of teaching with games, both in and out of the classroom, is to use the time spent in the classroom or doing homework to create a laboratory environment—an environment in which experiments can be made, hypotheses formulated, and new and better experiments planned. Games help to create this laboratory feeling by providing objectives and procedures. They also encourage imaginative freedom to experiment with alternative solutions, while at the same time offering a realistic set of constraints on less practical responses to problems. The students can learn not only by observing the results of games, but also by playing and indeed by designing them.

The first phase of game learning, the design and preparation stage, may be divided into two kinds of activities: relatively passive preparation for active game play, and the actual design of the game to be played. The former is likely to be more common, but the latter is probably more rewarding. The former involves simply learning the background material to be simulated in the game and the game rules. This is little different from conventional study, except that it tends to be highly motivated by virtue of the promise that one can express competitively and dramatically what one has learned, rather than merely regurgitating it in an examination.

In the latter, more rewarding way of game preparation, the game designer is actually inventing a simulation model of the process to be gamed. (This is also what social scientists, economists, engineers, and mathematicians do when they simulate a complex problem in more manipulable and simplified form for the purpose of staging experiments impractical in the real world.) In the course of doing so, the student must identify the significant variables involved, the relationships among them, and the dynamics of the interaction. To do this successfully is, in fact, to understand the process being simulated and to be able, in large degree, to predict its results accurately. Involving the students in this process expands their knowledge greatly: they learn not only factual content but also the processes, relationships, and interactions involved.

The most obvious way to teach with games is to play them in the classroom. This raises the practical issues of topic selection, timing, logistic arrangements, casting, materials, special requirements imposed on teachers, and interactions with other curriculum materials and activities. Most topics are suitable for educational gaming, some more so than others: topics that involve multiple forces or actors in some form of mutual competition with uncertain outcomes are most gameable. On the other hand, some topics which may not seem ideally suited to being gamed may be so remote from the students' interests that some form of classroom activity is required to stimulate interest, in which case games are a useful motivating technique.

Many required classroom topics and subjects are best learned by direct study—reading, observation, or field experimentation. There is hardly any subject, however, that does not have some interactive, competitive elements that are natural material for gaming—and other relatively static and formal elements that strain the mechanism of the game. To integrate these elements one must first of all divide the material to be learned into those parts with interactive and competitive elements and those of a formal or static nature. If, for example, the students are studying the Constitution of the United States, what the Constitution actually says is learned best from reading it. But the story of how the Constitution was written, why certain parts were written one way or another, what some of the alternative ways and interpretations were, can be well understood by means of a Constitutional Convention game, in which the competing political and economic forces can be dramatically simulated. The reading can be left for home study, with the essential interaction among individual and autonomous decision-making players being the only activity that requires the students to be together. After all, the classroom is the only place where students can directly interact, and this ought to be fully exploited by means of the teaching device that makes use of these interactions— namely, the game.

Games stimulate conventional study and can be used to summarize the results by dramatizing the interaction of disparate elements that were studied in isolation. If a given classroom study topic usually requires ten classroom hours, one could begin with a one-hour game and append to the tenth a one- or two-hour game—the first devoted to exploring the topic and the second to analyzing the results of study.

Most teachers who have used classroom games are enthusiastic about them, but some are ambivalent, and a few are violently opposed. The use of games for teaching, it has been argued, requires both too much and too little. At an educational conference in 1965, for example, a former head of the American Federation of Teachers suggested that educational games would both keep teachers too busy to do their jobs, and at the same time threaten their jobs by taking away their work!

A teacher using educational games becomes more of a research director and coach than a lecturer and disciplinarian. The game mode carries with it its own rules of behavior or discipline which must be observed by the participants if they are to enjoy the interaction of the game. The peers of whatever age involved in the game rarely break these rules because they know that it will end the game for them. Rarely do children in baseball or football games let the game fall apart because they cannot abide by the rules. Children do not want to play games they do not like, following somebody else's rules, but they are generally happy to play, and often insist on playing, games they like by the rules of the game. The teacher in a games classroom, then, need not be a disciplinarian. The time and energy saved may be directed to coaching the students to play the games more effectively, or observing their intellectual strengths and weaknesses when faced with specific types of situations.

These concepts are not new. Maria Montessori suggested over fifty years ago that the teacher spend more time observing students and less time directing them. More recently John Holt has reaffirmed this concept of teaching, which is not so much permissive as analytical—that is, teaching is seen as an activity of analyzing and responding to student performance rather than a constant attempt to control it. In this kind of education, the teacher has a responsibility to be much more analytical and lucid in the presentation of the game mode and in the analysis of the game consequences than is usually required by the lecture system. In a class where games are used, the teacher must learn to give brief but very intensive analyses and explanations, interspersed with longer periods of observations of student experiments and occasional coaching remarks. This is entirely different from the continuous pattern of doctrinaire topical material transferred from textbooks to the teacher's mind to the teacher's mouth to the students' pencils. And it should be more rewarding and entertaining. The teacher is now an attendant, an audience at a drama in which student actors display their problem-solving capabilities in intellectual contests. There is suspense over the uncertainty of the outcome, and emotional identification with contesting actors. But the teacher is also the critic, evaluating each player's interpretation of his role.

The teacher need not worry about being disengaged from the students in their play of serious educational games, for he is game director-referee-coach throughout, and afterward acts as chairman of a "de-briefing" or post-mortem game analysis. This post-game analysis should be a structured, directed discussion of the limitations and insights offered by the game and of the performance of the players in both representing and solving their problems effectively. Here the players will consider the teacher as a vital part of the game's operation and resolved meaning, rather than as a person who interferes with classroom activities.

Great demands are made on the teacher in understanding the processes being simulated and taught by the games, and in understanding and presenting fairly their most meaningful aspects. He must recognize the students' limitations in simplifying or neglecting some aspects of the processes simulated. But evaluation of students by means of the tiresome process of grading quizzes (which are often a better measure of how well a student writes than of how much he really knows) can be replaced at least in part by evaluation taking place *during* a game—evaluation by peers in the course of the game and by the teacher in both his observations and postgame analysis. Thus, a teacher's role becomes more demanding and interesting intellectually, and less demanding in routine clerical activity.

The timing of a classroom game should be made to maximize the game's dramatic impact on the students, either in terms of a culminating drama that weaves together diverse strands of one large topic, or as a way of introducing or making interesting and meaningful an otherwise abstract or uninteresting topic. Equally important is the timing of actions within the game itself. Here too the teacher-director has a crucial controlling function, for he has the power and, indeed, the responsibility to maintain the pace of the game so that maximal learning occurs. When the pace lags, he can often speed things up by introducing crises of various kinds. When players are indecisive or confused by too many trivial calculations, it is often effective to introduce a critical problem which demands immediate action from all the players or which affects their fortunes if action is not taken. In a political game, for example, the candidate teams contesting an election might be projected into more intensive activities if the referee announced that all campaign platforms must be presented on a simulated television program within the next few minutes. In an economics game, a temporary lag in buying and selling might be overcome by the referee's introducing a threat that affects them differentially, such as the loss of a given stockpile of goods or the sudden entry of a major buyer.

Disputes not covered by rules frequently arise in simulation games which allow liberally for individual players' imaginations. Where the interpretations of roles or parts in the game are left to the individuals, conflicting interpretations may arise. The good game director or referee will resolve these disputes as quickly and equitably—also as quietly—as possible. Interruption of the other activities of the game should be avoided by isolating the disputants from the rest of the game, resolving the crisis, or agreeing to resolve it after the game and requesting the players to go on with what they were doing. The author found that, in general, the best ground rule for settling unanticipated disputes is to resolve them on grounds that most closely correspond to the reality being simulated.

The scoring of a game is, of course, extraordinarily important. Players should be able to determine the relative effectiveness of their playing, who won or lost (or played realistically or unrealistically), and what effective play means in the particular process being simulated in the game. Scoring also provides what psychologists call "closure" to the activity, completing it in a psychologically satisfactory way.

Scoring can be done by a referee, the players themselves, or a combination of the two. In our experience, player scoring has been the most effective for educational and analytical purposes, because the details of effective and ineffective performance are closely scrutinized in the process. It requires all the players to have a clear and common understanding of the "win criteria" of the game, the penalties for violations of the rules, and the overall game purposes. In many complex simulation games, scoring cannot be reduced to a simple matter of assigning point values on the basis of a predetermined schedule to various types of behavior. It usually requires interpretation of the relative effectiveness of different moves both in terms of the realism with which they simulated the process under study and the effectiveness of the response in terms of the win criteria.

The scoring activity is related to but not identical with the evaluation of the game itself or of the players. The scoring is merely an indication of effectiveness. Thus, a player who makes many effective decisions in a game and scores high has demonstrated operationally his comprehension of the process being simulated, but the converse is not true. A player *not* effectively responding to the game situation in terms of the decisions he makes is not necessarily stupid or ignorant; he may have had something in mind that simply could not be expressed in the framework of the game. He may have believed that his ineffective responses were the most accurate possible representation of his role.

Similarly, games often identify intelligence, general problem-solving ability, dramatic talent, and negotiating talent in individuals who have difficulty demonstrating these capacities on ordinary intelligence and aptitude tests. But the absence of such talent in a simulation game is not necessarily significant; the player may have been unable to apply his particular skills in the context of the game.

In the course of game play, students will sometimes develop innovative and completely unanticipated solutions to problems. When this occurs, it is important that the teacher be flexible about game scoring. Usually a trade-off must be made—some students will consider the game unfair because it has not been scored completely by the predetermined criteria while others will feel that really creative thinking should be rewarded. Explaining the significance of the unexpected behavior and the reasons it deserves to be considered with the predetermined criteria is usually required.

Game performances, like dramatic performances, may be good, bad, or indifferent. The game content also, like the script of a play, may be evaluated as good, bad, or indifferent. If the game itself and the substance of it is given a low rating by the players, redesign and replay are obviously called for to better the simulation and approach the analytical objectives more closely. In this sense, the scoring of the entire game rather than of individual players is an essential part of the process that leads to a redesign and refinement of simulation games.

III

EDUCATIONAL GAMES
FOR THE PHYSICAL
AND SOCIAL SCIENCES

In MOST of the games we discussed in Chapter II, social interactions of all kinds were involved, and these almost invariably have at least some competitive, intrinsically gamelike component. Science, however, is man's great attempt to understand universal laws; if scientists compete with anything, they compete with nature. In the very human way of competition for prestige, they do, of course, compete against each other, but this is a game of one-upmanship, not a game of science.

Competition with nature is a misnomer, however, since competition implies some sort of comparability of competitors. When a weightlifter attempts to break the world's record, it is not meaningful to say that he competes with the force of gravity, even though he works against it. More accurately, he competes against other weightlifters in overcoming the force of gravity. All "showdown" or race games are of this sort—the players, though they compete against each other, "play" with some natural obstacles, such as gravity, energy limitations, friction, time, etc. So does the scientist "play" with natural rather than human obstacles.

This changes the nature of the game played, since most games usually involve a countermove by the opposition in response to a player's move. Unless one believes, as the ancient Greeks did, that there are malevolent natural deities, nature does not respond to a scientist's moves with conscious countermoves. So activities or "games" in which one pits oneself against a natural barrier (such as time, the force of gravity, or energy limitations) are in one sense simpler than games played against other players. The scientist or "player" need not be concerned with psychological warfare or with carefully calculated countermoves. On the other hand, the simplicity of interaction is usually counterbalanced by the difficulty of the objective: forming theories of relativity is not easier than becoming the ruler of a great nation.

A second important difference between games for the social studies and games for the sciences is in their motivational objectives. In a classroom simulation of an economic, political, or social process, the players can participate in the process. They can "become" members of the Supreme Court, leaders of warring nations, or Congressmen voting on important national legislation. In the science classroom, however, the student usually simulates the activity of the scientist: he conducts experiments or develops theories and solves theoretical problems. The abstract nature of the subject matter itself is not distorted or made more abstract by this static presentation; and role-playing would be at best an artificial, contrived technique.

The chief contribution of games to the study of sciences is the expanded scope their use affords. Evolution, entropy, relativity, astronomical relationships, and quantum mechanics, for example, are not readily demonstrable in elementary- or secondary-school classrooms. Consequently, these seminal ideas are often either omitted entirely or presented in much the same static and taxonomic way that social-studies material is presented in conventional classrooms. This material, however, can be made vivid and memorable when games are employed to simulate the development or operation of these concepts. Students do not role-play or simulate human interactions: each student "plays" against nature to the best of his ability, the winner of the game being the student whose understanding of nature has been most complete. Since these natural forces or laws do not change, students can test alternative courses of action against them and can evaluate their results.

Topics already included in the science curriculum may also benefit from simulation games designed to motivate the student to go beyond the required study materials. Many students, particularly elementary schoolchildren, regard science and mathematics as problems in themselves, rather than as instruments for solving other problems. But numbers, things, the look and feel and smell and sound of things and how they move, are themselves objects of study, although the students may not be consciously aware of any "scientific method" in their exploration of them. But when problems related to the real world are made attractive enough through simulations, most students will quite naturally use all the methodological and technical resources in their capacity to try to solve them.

Perhaps we should not limit our purview to pure simulation games but should also consider interactional simulations that are not really games in the sense of winning and losing, but which are exercises or projects with a certain dramatic interest for the pupils.

In the first three grades of elementary school, instruction in the sciences should ideally be distributed among general problem-solving logic (heuristics), computation (arithmetic), and study of the world (physics, chemistry, biology, geology, etc.). Conventionally, the last two are taught separately as "arithmetic" and "science" and the first is not taught formally at all but is learned while playing, when rote answers to operational problems are scorned and leadership is asserted at least partly by innovation in technique.

Many children have difficulty mastering the abstractions of numbers and arithmetic, the imprecision of nonquantitative presentations of the physical world, and the presentation of "solutions" to be memorized without meaning. "Two of *what* plus three of *what* makes five of *what?*" the child's mind wonders. Or "How can the sun be further away from the earth than the moon, when it looks about the same size?" In response to this problem, Montessori and others developed multisensory physical representations of arithmetic abstractions, such as Cuisiniere rods. It seems an almost obvious step to combine arithmetic and science so that quantities are related to real things, and real things are regarded quantitatively—that is, with precision, objectivity, and calculation. One promising interdisciplinary approach to combining elementary arithmetic and science is to use models or simulations of natural forms and events. These simulations may be exercised by students individually "against nature" as in a puzzle, or interactively as "games."

The dramatic psychology of games is important to apply even to the individual puzzle exercises. A puzzle is no fun if one knows the answer or if it is impossible to solve. The games should stress the student's own discovery of heuristics, computational techniques, and scientific facts as instrumental to his objectives in a contest with nature or with other students. In an elementary science game called "Forecast," for example, a contest in making predictions about temperature and precipitation can be based on arithmetic computations of the movement of a cold front, incorporating scientific facts, quantitative analysis, and logical inferences. Students can be given a map with a variety of locations, each reporting its local weather on the basis of a series of plastic overlays keyed to the map, on which local conditions are printed. And so on.

Other games for elementary and preschool science can stress an interdisciplinary approach. In a game called "Weather," the aim is to enable the student to identify and prognosticate about a variety of weather states. The object of the game is to select the correct clothing and activity on the basis of visual weather data. A series of photographs is shown, each depicting a different weather state in alternate morning and afternoon pairs. The student players are shown the picture of the morning conditions and must then choose an activity and clothing for the afternoon. The afternoon's weather picture is then shown, and players are scored according to their plans. Those who chose activities and clothing appropriate to the afternoon's weather win points, and those whose plans were ruined by the weather are penalized. The winning player has the most points after a series of these sequences.

The educational aim of a game called "Color" is to enable students to compose colors from the primary colors. The players are given a transparent, gridded plastic game board in which some of the grid positions are red, yellow, and blue, with most positions blank. They are also given transparent plastic markers the size of two grids. The players are then divided into three teams: Orange, Green, and Violet. The Orange team has red and yellow markers, the Green team has yellow and blue, and the Violet team has blue and red markers. The teams take turns laying down markers on the plastic grid to create chains of their team colors all the way across the board. The markers must overlap to create the right color. One team can preempt or block another team by overlaying the other teams' markers with its own. The first team to make its own color chain cross the game board wins.

In "Shape," the educational objective is to enable the student to distinguish and construct a variety of common geometric forms. The game is to recognize those forms that occur in game context and their combination into new forms. The only equipment is a game board with pictures on it of common objects representing the geometric forms of a sphere, disk, cylinder, and cylindrical shell; and variously sized hoops with straight and curved sides for constructing cylinders and spheres respectively.

A supply of hoops is dealt randomly to players, who in turn place one of them to a picture, successively adding hoops to form the pictured object. Players can add to each other's constructions or initiate their own. The player completing a form gets credit for it, with the most points being awarded for the most complex form (sphere) and the fewest points for the least complex form (disk). Players may also challenge a "completion" by adding yet another hoop that does not change the form. The player gaining the greatest number of points is the winner.

The study of mathematics alone can also be enriched by the use of games. After all, children outside school spend hours on card games requiring mathematical operations, strategy, or conceptualization. In the classroom itself, using the students' enormous energy in a positive way—in games—greatly reduces discipline problems, lack of attention, and failure to participate, which often occur when arithmetic is taught by drill. It does not, however, reduce the noise level of the class. When students play arithmetical games, they should be required to recite aloud the arithmetical processes they use. They are also involved in competing to win the game and tend to argue with one another about the correctness of particular moves. It is precisely this spirit of competition which one should exploit to engage the student in the learning process.

Games also reinforce the skills and concepts mastered. Once a student is in a game, he is unlikely to withdraw from the other participants or to ignore the moves they make. His understanding of the principles involved is therefore reinforced not only in what he does but also in what his opponent and/or partner does. In addition, the incorporation of error-checking moves, whereby one student gains an advantage over another by identifying and correcting an error, will contribute to learning. And cooperation among team members encourages the better student to help his partners—individual help that the partner might not otherwise get.

Students should not have to play one variation of one game

for an entire classroom period; there are enough games to provide for great variety and for review of many specific processes. Because the chief objectives of the games are drill in arithmetical principles and conceptualization of mathematical problems, it is best if pencils and paper are not allowed. (The more difficult games, however, may frustrate the student who is not completely ready for them; in this case, he ought to be allowed to use pencil and paper rather than made to try an easier variation of the game.)

The deck of cards used in one set of math games we have devised is specially constructed to allow for the greatest freedom in making moves, to have enough cards of a given number value to make problems generated at random solvable, to allow for rapid play, and to introduce enough elements of chance to be interesting and competitive. There are sixteen cards for each number value from 0 to 9, and eight "wild cards" which can represent any number from 0 to 9.

"On the Button" is the simplest of the six games played with this deck.* It requires at least two players. The goal is to reach a predetermined target number by using cards dealt in a hand and drawn from a pile. The player who reaches the target number first wins all the cards played in that round. Play continues until all cards in the deck have been used. The player who wins the most cards wins.

Here is a sample game: Two players agree that the target number shall be 20. Each is dealt two cards from the deck, and the rest of the deck is placed face down. Player A plays a 3, saying "No cards plus 3 equals 3." He then draws from the deck to replace the card he has just played. Player B plays a 6, saying "3 plus 6 equals 9," and draws a new card. Player A plays an 8, saying "9 plus 8 equals 17," and draws. If he can, Player B will play a 3, saying "17 plus 3 equals 20"—the target number. He then takes all cards played in that round and draws. Player A then

* The remaining five games are included as Appendix A.

starts a new round by playing a card from his hand. Play continues until all cards in the deck have been played. The winner is the one with the majority of cards.

Obviously, it will not be possible to reach 20 exactly on each round by using addition only. When the total reaches twelve or higher, the next player might have to play a card which would bring the total above the target of 20. All cards from that round then remain on the board and a new round is started, until one player in one round reaches 20; this player wins all the cards played in the completed round plus all cards for incomplete rounds.

If a player makes a mistake, the other player may challenge. If the challenge is correct, the challenger wins all the cards on the board; if incorrect, the challenged player wins the cards. This "mistake policy" insures that each player will check the other and that players will think before speaking.

There are numerous variables which the teacher can manipulate to tailor the game for the desired degree of difficulty: the number of players; the number of cards each player will have in his hand; the target number; and the number of arithmetic processes (addition, subtraction, etc.) to be used in the game.

Turning from the science curriculum to social studies, one finds that in most elementary and high schools and in most junior colleges, they are taught more like the humanities than the sciences. Descriptions tend to be literary and qualitative, predictions based on judgment rather than on empirically validated analysis. Experimentation is rare and imperfect, the few attempts usually marred by an aggregation of uncontrollable variables. In most social studies courses there are few convincing formulas explained, no theoretical computations made, no experiments executed or analyzed. Social studies are usually taught as a literature devoid of dramatic characterization, explanations that fail to satisfy or have predictive power—evaluations more personal than persuasive.

Much of learning is based on experience and making associations between present problems and past experience. Talking, listening, and reading provide a direct experience of the content of literature courses, as do measuring and estimating in mathematics and experimenting and predicting in physics and chemistry. But in social studies, as conventionally taught, few if any experiences of what is to be learned are made available. The students have no chance to experience or experiment with social, economic, and political issues, to make or write history, to solve problems of government, economics, and social organization.

One major flaw due to this lack of experience is the absence of memorable surprise—the sudden exposure of error or success. The differences between truth and error are not demonstrated experimentally, but preached on the basis of some impersonal and often unconvincing authority. Students cannot make mistakes: they can only fail to memorize someone else's mistakes or successes. But students cannot learn to correct their mistakes without being allowed to make them. And making mistakes in government or economics means making bad practical decisions; it does not mean mistaking a name or a date.

The valuable educational experience of decision-making in various social, political, and economic roles can be provided in the real world only at great risk or cost. Boys' Day at City Hall is not really such an experience unless the city is willing to trust its real decisions to the students and let them see the consequences. The only alternatives to the impractically costly direct experience are abstract and diluted ingestion of other people's experiences, or simulated experiences.

Such experience in social, political, and economic decision-making can be gained when students participate in simulations of the real-world processes. The major problem is to design simulation games which imitate reality well enough to be used, yet which are sufficiently simple to be played by classroom-size groups in short time spans without elaborate equipment.

The table on pages 44 and 45 shows some of the elements common to both social-studies topics and formal games, providing a basis for the design of educational simulation games.

Now, the study of war and peace is of course a major "social"

SOCIAL STUDIES		ELEMENTS COMMON TO SOCIAL STUDIES AND TO GAMES			FORMAL GAMES	
Subject	Topic	Players	Objectives	Typical Resources	Example	Type
History	Civil War	Loyalists vs. Rebels in Civil War	Gain support of neutrals	Military power, propaganda		Strategy (sequential player actions that are mutually responsive and stress need for predicting opponent's moves)
		High, low vs. middle in High-low poker	Gain support of opposite	Card suites, betting (persuasive) strategy	High-low poker (2 winners)	
	Colonization	Colonizers vs. Colonizers	Control colonial region	Power, decision, speed, determination		Showdown (simultaneous player actions depending little on uncertainties of opponent's moves)
		Climbers vs. Climbers	Control "mountain"	Power, decision, speed, determination	King of the Mountain	
Geography	Raw Materials Production	Producer vs. Producer	Capture market	Location closest to market		Strategy
		Player vs. Player	Capture ball	Closest to ball	Soccer	
	Trade Routes	Civilization vs. Geography	Get closest to market	Mobility		Showdown
		Players vs. Position	Get closest to objective	Movement	Shuffle-board	

	Legislative Processes	Elected Reps. vs. Elected Reps.	Vote or kill legislation in spite of blocs	Votes on right issue at right timing		Strategy
Civics		Players	Score goals, deny opponents goals	Players at right place at right time	Basketball, football	Strategy
	Elections	Candidates vs. Candidates	Win	Outdistancing opponents		Showdown
		Racers vs. Racers	Win	Run faster	Races	Showdown
	Union-Management Collective Bargaining	Union vs. Management	Increased share of profits	Strike vs. lockout		Strategy
		Teams	Goals	Massed power	Rugby	Strategy
Economics	Competitive Investment	Investors vs. Investors	Profit	Capital, calculation, luck		Showdown
		Players	"Play" profit	Capital, calculation, luck	"Monopoly"	Showdown

studies problem, since issues of war and peace are a critical concern of national governments, and the study of national governments is an essential part of studies such as history, political science, economics, and sometimes anthropology. If war and peace are proper subjects for social studies, so are "war games." War games have been played for centuries by soldiers and civilians—sometimes for education, as in the Prussian war games of the early 1800s, and sometimes for pleasure, as in chess.

War games usually involve interactions of adversary forces on maps representing arenas of battle. Sand tables were sometimes used in the early days to represent deployments of troops. As real wars became larger in scope and more complex technologically, map representations and computing devices in games were added. Increasingly, conventional "war games" became too narrow a simulation of a diversified spectrum of international conflict. Since the development of nuclear weapons, the unprecedented threat of near-annihilation of nations has meant that national and ideological conflicts of less destructive kinds—such as economic, political, and propaganda warfare and subversion —have become ever more important; these too can be simulated by games, as are "limited wars," including internal revolutionary conflicts, insurgencies, and coups d'état.

Let us consider as an example of a modern "war game" one of the most complicated games ever devised for the social studies. In 1961 the Raytheon Company was asked by the Joint War Games Agency (JWGA) of the Joint Chiefs of Staff to develop a computer simulation model (a computerized war game) of global cold-war conflict. Their simulation model was unique in its scope and complexity.[*] Three power blocs, or alliances, later thirty-nine nations with conflicting interests, confronted each other in economic, military, and political modes over ten-year

[*] See "Strategic Model Simulation of Global Military, Economic, and Political Interactions," Raytheon Missile and Space Division, Bedford, Massachusetts, BR-1795, June 1962; and "Design for a Strategic Model," Clark C. Abt, Raytheon Company, Bedford, BR-1354A, 24 September 1961.

periods broken down into weekly events. As in the real world, but as opposed to the world of traditional "war games," all sides could win long life and prosperity by practicing arms control, or could lose life and wealth in unrestrained arms races or aggression.

The computer program, which consisted of approximately twenty thousand mathematical instructions, failed, however, to give some of the users and designers a clear idea of the dynamics of the model, even with numerous briefings. A check on the plausibility of the model was also needed, in a form accessible to the users unskilled in computer language.

To clarify the model and to educate both users and designers in applications of contemporary theories about alliances, deterrence, economic development, decision-making, resource allocation, etc., a game was designed to simulate the computer model in more elementary, accessible, and dramatic form. The game was called "Grand Strategy" (not to be confused with the secondary-school game of the same name described in Chapter II). The three geopolitical blocs were represented on a gridded four-by-eight-foot map of the world; elements of the computer model called subroutines were translated in somewhat simplified form into game rules.* Computer submodels representing the national political, military, and economic decision-making groups were translated into player roles as political, military, and economic leaders for each nation. So this "manual" (as opposed to computer-operated) game was an analogue of the computer simulation, which in turn was an analogue of international relations as shown on page 48.

Originally, "Grand Strategy" was played by three teams, representing the Eastern (Communist), Western, and neutral geopolitical blocs. Each team had a military player, who moved military pieces, and an economic player, who moved economic chips, both under the direction of a political player who coordinated over-all bloc policy and conducted negotiations.

It soon became obvious that blocs were not really monolithic

* The model is no longer maintained by the JWGA.

organizations; and if realism was to be preserved, then the game had to be modified to take into account internal divisions over economic interests and ideologies, and breakdowns of communication and coordination. The bloc teams were split into loosely coordinated nation teams. When there were enough players, each nation team had three players; when only a few players were available, one player would represent all functions in a nation.

International Relations	Computer Simulation Model	Simulation Game
Planning	Central decision-making submodel	Political players, word of mouth and secret notes
Information-gathering	Diplomatic submodel	"
Negotiation	Bargaining submodel	"
Trade and Aid	Economic submodel	Economic players, chips
Military affairs	Military submodel	Military players, forces and weapons pieces
Culture and ideology, international friendship and hostility	Cultural-ideological submodel	Political alignment track representing ideological distance
Technology and science	Scientific submodel	None, but provision for technological change in weapons and production efficiency.
Territory	Area code	Map board

Originally, tables of intra-alliance and inter-alliance solidarity were used to express the probability of alliance formation and dissolution under a variety of circumstances. After "Grand Strategy" was played a few times, these tables were found to be unwieldy, and a visually more direct and dramatic expression of alliance shifts was called for. A "race track" in the form of a horse-shoe was developed to represent relative political alignment in terms of graphic distance:

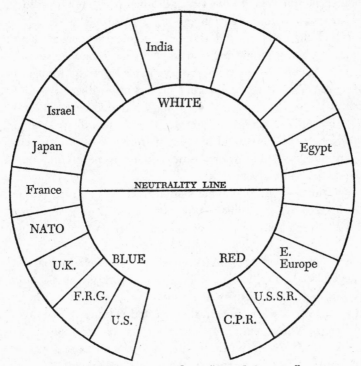

Political Alignment Track in "Grand Strategy"

The purpose was to resist being moved around the track against one's will, while at the same time to induce other nation planners to move as close to one's own part of the track as possible. The proximity of the neutrals to one end or the other depended on their political-economic sympathy or antipathy to the two blocs. Nations could move in either direction along the track, depending on the actions of the other nations and how they were perceived. It was possible to induce an ally in an opposing bloc to become neutral, and then possible to induce him to join one's own alliance.

The major goal of the game was to form an alliance which remained united, close to one's own original position, attracted the greatest number of neutral and enemy nations to itself, and amassed the military and economic resources to continue the process while avoiding major wars. National economic resources could be invested in domestic economic growth, foreign aid (economic growth of an actual or potential ally), or military forces. All of these investments could be traded in again for liquid economic assets, at some cost. For example, chips signifying production facilities could be bought from the bank by a nation team and "invested" by depositing them in one of the country tracks on a "status board."

On this second board, each nation track was divided into economic-growth and military-capability bins. During each simulated year of play, all the economic chips on the board were shifted forward one bin, maturing into the nation's intended economic production or military capacity after four annual shifts (an arbitrary simplifying assumption). Nations giving aid or economic investment (aid with a hoped-for profit) to another nation "invested" their chips in the invested nation's track in return for some desired military, political, or economic action.

Countries moved on the political-alignment track toward the bloc from which they received the most economic aid by a number of places determined by the number of a rolled die multiplied by the ratio of the aid from the two competing blocs. If, for example, the West (Blue) built four factories in Africa and the East (Red) two, Africa moved six spaces toward West at a roll of a three on the die $(4/2 \times 3 = 6)$. The uncertainties introduced by the dice were intended to represent the resolute independence of proud and poor nations.

Military moves consisted of production and deployment of forces, and conventional and nuclear attack. A major limitation in the game was that insurgency or internal guerrilla attack, at least equally significant, was omitted for the sake of simplicity; but this could readily have been added. An attack on one or more members of an alliance constituted an attack on all members of the alliance. Members who moved away from their original positions on the political-alignment track returned, to simulate the drawing together of weak allies under a common external threat. In a nuclear attack, however, neutrals and weak allies could retreat further into neutrality. Combat results were decided by means of predetermined tables. Given the ratio of one's own to enemy forces and levels of attack, the tables gave the results in terms of forces and economic productive power destroyed on each side.

When the game was test-played by its designers, military systems analysts, and Pentagon officials, the players showed a surprising identification with the national styles and aims of their respective roles. They became involved in the game to such an extent that game losses were felt as personal losses. Arms reductions by both Eastern and Western nations were accomplished unilaterally or through bargaining—perhaps in part due to the liberal views of some of the players, and because of the practical incentive of developing sufficient economic power to woo the neutral bloc with foreign aid investment. When the neutrals rewarded aid with political solidarity, international stability and arms reduction occurred. But when the dice rolled unfavorably and neutrals offered little or no return for investment, incentives for stability were weakened. Also, with the West's certainty of successful alliances as a result of aid investment, came corresponding certainty of political losses to the East. When the situation became hopeless for the East, as it did on two occasions, the Eastern nations resorted to nuclear war with a "back to the wall" psychology. This irrational behavior tended to accumulate. Then, as tacit and explicit arms controls on strategic forces were implemented, due to the apparent unfeasibility of "safe" surprise attack, the resources saved by reducing strategic nuclear forces went into conventional forces; the result was many conventional limited wars.

A significant change occurred in the players' understanding of international relations. From the presumption that the principal conflict in the world was strictly military, bipolar, and purely competitive, the participants came to realize that it had equally if not more important political and economic dimensions, was multipolar, and was a mixture of cooperation and conflict. Since the theoretical judgments, beliefs, and factual knowledge of the game's designers limited the possible outcomes of the games, little new knowledge about the real world emerged from the gaming, but new comprehension was gained when the players saw the consequences of alternate arrangements of the known facts.

Many of the techniques developed in the design of "Grand Strategy" are applicable to the design of less complex simulations. The major difference between games for secondary students and games such as "Grand Strategy" for adults is that limits on time and cost also limit the possible complexity of the games for students—but almost never their cognitive skills. It has been our experience that the average twelve-year-old can perform all the logical analysis of the problem that an average adult can. So the secondary-school student is inferior to the adult in problem-solving power only in having less information and experience—qualities educational games are supposed to supply rather than demand.

Given this assumption about the intellectual capacities of high-school students, the chief concerns of educational game design at this level are to pack the classroom hour (or a more suitable unit of time, if scheduling flexibility and administrators' imaginations permit) with as profound and richly detailed an exercise of the problem under study as possible, to maintain student interest, and to dramatize the primary questions with sufficient contrast to secondary ones.

Many edvcational simulation games designed for secondary school use are equally suitable for college and adult use. The only thing that makes them "high-school games" is that they were designed as such. But they are not limited to schools and might just as well be used for teacher-training in the particular substantive areas covered. In fact, adults sometimes are outperformed in these games by high-school students. This should not be a source of anxiety, however. The students are more habituated to games and probably less inhibited by concern over the impression they will make on their fellow players. They are also less compulsive about getting all the details exactly right, so that they often are relaxed enough to see the over-all situation more quickly.

As an example, the game of "Empire" was developed by the author for junior-high-school students when it became apparent that most contemporary American students could not understand the degree of the colonists' outrage when Britain enforced the Navigation Acts. The game simulated eighteenth-century trade among England, Europe, and the various American colonies. The player roles were London merchants, New England merchants, colonial farmers, Southern planters, West Indian planters, European merchants, and the Royal Navy—most of the major decision-making groups in the great mercantilist struggle between England and the colonies in the mid-eighteenth century.

The early players of "Empire" seemed to understand the game very well. Although the ideological bases of conflict between England and the American colonies were not made explicit, three little Negro girls, when asked what roles they preferred, immediately chose the London merchants: they quickly understood that they were the creditors of the Southern planters.

Classroom experience with "Empire" revealed a major paradoxical problem that occurs in all social-studies games: the conflict between a "fair" game and a historically accurate simulation. Because students like to win, they tend to object to situations in which teams do not all have an equal chance of winning. Yet in most real historical, social, economic, and political situations, the "teams" or "players" do not in fact have equal chances of winning their objectives. Not even the God of the Old or the New Testament can claim that the world is a fair game. (Job and Jesus suffered punishments they themselves did not "deserve.")

The alternatives were to distort the simulation of reality to achieve a "fair" game, or to accept "unfairness" for the sake of scholarly verisimilitude, or to devise some compromise of the two. One way of doing the latter was to build in handicaps or to measure team gains in terms of relative rather than absolute increases. In "Empire," for example, although the colonial farmers' team begins with much less wealth than the New England merchants, they may still beat them by gaining a greater *percentage* increase in their own wealth. Sometimes, children complained that too many of their ships were seized by the Royal Navy (for smuggling or pirating), despite the frequency of such events in history. To some extent this "unfairness" also introduced dramatic interest and was suffered equally by several (if not all) teams. Or again, because mercantilism enforced by the Navigation Acts favored the London merchants, this team was favored in "Empire." This "unfairness" did not eliminate play incentives, however, because it could be circumvented by smuggling (children delight in simulating illegal acts!), any team could still "beat the system," and the scoring was based on proportional rather than absolute gains to equalize initial inequalities.

A few teachers have raised a moral issue concerning these games—not whether games are too frivolous an activity for the classroom, but whether games "teach" immoral behavior. In "Empire," the historical immoralities simulated include slave trading, smuggling, and bribing customs officials. Some teachers felt that the simulation of these activities by students implied the condoning or even the encouragement of immorality. One high-school principal even said he felt it was unpatriotic to represent the colonists as being engaged in such activities. The fundamental issue here is whether social studies should deal with reality descriptively or in ideal terms or both. If social studies ever hope to be social science rather than moralistic indoctrination, they must present reality as it is and was, unfiltered by values. To do so does not imply abdication of moral values any more than a medical doctor's study of the damage done by a disease implies that he accepts the disease as desirable. And it is important to remember that "Empire" was developed for children of twelve to fourteen years of age—an age group that can well deal with abstractions and questions of morality and seems to enjoy doing so. The immoral historical activities simulated provide the teacher an opportunity to discuss the moral questions with the students in a realistic context.

Very young children, however, tend to see things in black and white, as "right" or "wrong." Anyone who has tried to explain to a six-year-old the difference between "stealing" and "borrowing without asking," or why punching is not just retribution for someone who has borrowed a toy without asking, is no doubt familiar with the difficulty of moral discussion with a six-year-old. Raising complex questions of morality in a game for six-year-olds would be equally futile.

Particularly in high school, students are often estranged from school by the refusal of some teachers to deal with real problems and issues. To many students, school has become an artificial activity having little relevance to the real world—a world in which many immoral and otherwise harsh realities must be dealt with, as the students well know. Social-studies simulation games should, insofar as possible, continue to describe events as they are, rather than as we wish they were.

The game techniques of "Grand Strategy" and "Empire" can be still further simplified and used to develop elementary-school games, where the level of abstraction must correspond to the more concrete type of thinking common in children from six to ten. Concrete physical activities are simulated, rather than the already partially abstract activities of, say, voting, budget allocating, and long-range planning.

A "Hunting Game" was designed initially to show how the African bush environment forced the Kung bushmen to hunt in socially organized bands, rather than individually. (Most of the prey was so large that not all of it could be eaten or carried by a single hunter or even two hunters.) Players representing hunters were given pieces to move a fixed number of squares on the gridded game board in the simulated search for prey. As a player placed his piece on a particular square, he would turn over a card keyed to it indicating the sighting or absence of an animal. (These cards were arranged randomly and secretly.) After sighting an animal, the hunter could elect to attack or, in the case of dangerous animals and modest hunger, retreat. If he decided to attack, the player rolled dice and entered the numbers rolled and the number of hunters on a table that determined whether the prey was killed, was wounded and to be tracked further, escaped, or killed the hunter. This table was based on different probabilities of kill for different animals—naturally the probability of kill increased with more hunters. The hunters had only a fixed number of turns they could move without the sustenance of animal food. As they "starved," their movement allowance was reduced until they became immobilized. If help did not arrive in the form of another hunter with meat to share, the starved hunter "died" and was out of the game.

The educational objective of this original "Hunting Game" was to teach the students the necessity for cooperation in the social organization of the Kung bushmen. By hunting in small groups, the bushmen increase their safety and their probability of killing any animals they encounter and can more fully exploit the kill by eating more of it.

Students in a Newton, Massachusetts, elementary school who played the game several times enjoyed it and soon learned to organize into small bands for the sake of survival. In addition to learning the need for organization for gathering food, the students also gained a knowledge of typical fauna of the area, ecology, hunting weapons, and negotiations for organizing and sharing. The game demonstrated that children as young as nine could with minimal supervision engage in an educational simulation game.

A later form of the "Hunting Game" was called "Bushman Exploring and Gathering"—this time concerning cultural adaptation to the grueling physical environment of the Kalahari Desert. In the "Exploring" phase, the players act as small groups of bushmen exploring an uncharted desert for food and water.

They do this by moving pieces on a large gridded game board with stylistically represented geographic features, inconspicuous water holes, and "hidden" animals. As the pieces land on a new square, players are allowed to turn over a card describing details of the local area which are not observable from adjacent squares. They then select and record whatever information they think is useful to their search and to their subsequent move.

Restricted in how far they can move on the board by water supplies, the players search for environmental clues indicating water and edible flora and fauna. They can increase the efficiency of this process by division of labor, by organizing themselves into non-overlapping search groups, and by recording where they have been and what they have found. This elementary form of the game, which takes about an hour to play, serves to familiarize the players with the search procedures and environmental variations that must be understood to solve the problems of the second, "Gathering" phase. They also learn to invent and apply a symbol system for recording the information they discover.

In the "Gathering" phase of the game, the objectives are to find the scarce food and to survive, toward which ends exploring is only one means. This phase also takes one classroom hour. Chance cards—which bear information about weather, and the relation between family size and water supply—limit movement. Families must obtain food supplies or leave the game. (The factors in the game are arranged to threaten but rarely to result in starvation because players excluded from the game do not learn much by being excluded.)

The performance of several groups of students who have played both phases of the game was evaluated on the basis of a comparison of written tests before and after the game. The evaluation seemed to indicate greater learning of cultural adaptation concepts than is usually attained by the standard (lecture) presentations in the same amount of time.

Two additional forms of the "Hunting Game" were developed to explore the relationship between technology and social organization in Eskimo culture. In the "Caribou Hunting Game"—a simulation of hunting strategies—several independent groups, each with three players, are seated around a desk-size board representing some ten square miles of land and water near Pelly Bay in northern Canada. Two students take the roles of Eskimos attempting to intercept a herd of caribou moved by the third player. The hunters are slower than the caribou on land, this being simulated by a smaller per-turn movement allowance. However, the hunters gliding over the water in kayaks can move faster than the swimming caribou. The hunters can try to kill the caribou with bows and arrows or drive them into the water to spear them from kayaks. The hunters must learn to cooperate and make maximum use of their weapons and transportation technology to gain their objectives.

The "Seal Hunting Game" should follow the study of material in the caribou game. Six to ten fifth graders (or fourth or sixth graders) gather around a board representing the polar ice covering the Arctic waters of Pelly Bay. In the board there are over a hundred removable corks representing the breathing holes of seals. The student players "fish" for seals by selecting a breathing hole under which a seal will be shown if the cork is removed. The players experiment with the hunting and food-sharing strategies needed to deal with the unpredictable food supply—division of labor, food sharing, and bargaining.

These hunting games all simulate the impact of rigorous environments on so-called "primitive" cultures. The relatively exotic and primitive scenarios and settings are intended to put into high relief the mutual impact of physical environment and human behavior—something more difficult to do with contemporary levels of technology.

Even in the first and second grades, social studies can make effective use of role-playing games. Such games seem to be most appropriate at this age, and it is amazing just how quickly the children accept roles and identify with them—even "roles" of inanimate (although not static) objects. The game of "Transportation," for example, has first and second graders assume the roles of buses, trains, ships, and airplanes. The children are assigned places of origin and destination objectives, fixed amounts of money, and simple tables of transportation costs and rates of movement. They are placed around the four sides of the class-room and asked to get to their designated destinations as quickly and/or cheaply as possible, imitating the sounds and movements of the vehicular mode they choose as they go. The flapping of wings, salivating splutter of motors, whine of buses, and hoarse cries of foghorns are a little noisy, but at least at the Boardman School in Roxbury, Massachusetts, in 1965, the children really identified with those vehicles and got to where they were going as quickly as possible.

The relative speeds and costs of alternative transportation modes, as well as considerable arithmetic computation and geography, are some of the things learned. The teacher may arrange for the game's emphasis to be on map locations or cost and schedule computations, depending on whether the main instructional objective is geography or arithmetic.

Other social-studies games can be developed to illustrate decision-making in historical, social, political, and economic roles, and they can be designed for any type of audience. As can be seen with the examples in this chapter, many social studies subjects can be made to overlap in a single game.

For all of these games, from "Grand Strategy" to "Transportation," the primary requirements of games have been observed. All are predicated on the active involvement of participants seeking to achieve objectives in some limiting context. They illustrate dramatically real problems in real contexts and provide for the exploration of alternative strategies. They provide for communication of ideas, many of which are abstract, within the familiar context of role-playing or competition. And, equally important, they offer the participants a means of expanding their own knowledge through vicarious participation in experiences normally reserved for others.

The student involvement with games, so much higher than with the conventional "read-discuss-write" technique, prompted one student to say, "You can learn things by reading, but game-learning is more fun, and what have you got against making it fun to learn?"

IV

GAME LEARNING
AND DISADVANTAGED
GROUPS

A SOCIAL PROBLEM receiving a great deal of attention today is that of educating students from "culturally deprived" or "culturally disadvantaged" areas—inaccurate euphemisms for the slum or ghetto. (The allegedly "culturally deprived" are actually only culturally *different* from persons patronizingly using the adjective "deprived.") Because this problem is such an overwhelming concern both to educators and to socially conscious citizens everywhere, and because games have an immediately useful application to the problem, it deserves some special mention in this book.

The problems of ghetto residents tend to be self-perpetuating unless, through education, the children of the ghetto can break the cycle of poverty which threatens them. But these children face great obstacles when they enter the educational world. Many problems are those they have brought with them; others they find ready-made on their arrival.

Let us examine some of these problems at their worst, recognizing that not all problems exist for all students in all school systems and that there are exceptions to all generalizations.

In most middle-income, suburban families, both parents tend to encourage the school activities of their children. In lower-income families, often only one parent is present (usually the mother), and this solitary parent is generally too preoccupied with meeting the physical needs of the children to spend much time and energy on their intellectual development. If the mother works, the older children may have to go to school in rotation so that someone will be home to care for the preschool-age children. In cases where both parents are present, the mother is often the dominant figure, particularly where the father is looked upon as an inadequate provider. (In Puerto Rican families, however, the father is usually very dominant.)

In low-income, low-education families, parents often behave inconsistently, meting out quick physical punishment for certain behavior at some points, ignoring it at others. The parents spend little time explaining *why* the child must behave in a certain way—they want complete obedience in areas which concern them at the time, so they can carry out the business of the family as efficiently as possible. Areas of behavior which do not affect the parents directly, such as table manners, personal hygiene, and school attendance and achievement, are often ignored as irrelevant.

Lower-income families, many of them with four or more children, live in small apartments with limited privacy. Study space is a scarce commodity, and it is not unusual to find two children using an ironing board as a desk.

The schools are often perceived as foreign institutions by children from low-income ghetto areas. Their curriculum, chosen by administrators of different cultural background, have only very indirect relevance to their daily lives. Except for basic learning, there is little carry-over to the home, where parents are often suspicious of threats to their authority. The teachers are middle-class, live in suburban or "better" areas, and have middle-class expectations—often in those very areas of behavior which the parents think are irrelevant. The teachers' attitudes and disapproval create conflicts in the children, who do not understand what they have done wrong. Because the home has taught the children not to ask *why* to questions of behavior, teachers often do not realize that what they expect is completely foreign to the students' experiences. If the teachers attempt to explain *whys*, the levels of abstraction they use are often incomprehensible to the students.

Due in part to their lack of experience with abstractions, these students find it difficult to sit still and just listen to a lesson. They tend to respond to the (to them) boring material with disruptive physical activity and low attention spans—which often causes teachers to regard them as "slow learners" and/or discipline problems. (American Indian and other poor *rural* schoolchildren, on the other hand, sit very quietly but also learn poorly.)

Because the teacher cannot by himself make the book-learning carry over completely to the daily lives of the students; because he cannot get them to understand what he considers to be conventional behavior; and because students give only brief attention to boring subjects and fall behind in "regular" classes, the teachers become understandably discouraged. A recent report indicates that about half of the teachers of slum children were not happy in their jobs, and that the median number of years a teacher spent in slum areas was only four, as opposed to nineteen years in the "better" areas.* It is also true that many white middle-class teachers dislike the students, consider them unteachable, and feel that their careers are foundering in the back waters of ghetto schools, which are not "good places to get ahead." Unfortunately, the teacher's dissatisfaction with himself

* Robert J. Havinghurst, *The Public Schools of Chicago: A Survey for the Board of Education of the City of Chicago* (Chicago: The Board of Education of the City of Chicago, 1964), p. 346.

and with his position creates a tension in the classroom which the students cannot but feel, yet cannot understand.

In addition to the problems of teacher-student communication, the schools also have problems caused by the condition of the buildings themselves. The ghettos of the poor tend to be in the oldest areas of a city—areas which have been exposed to successive waves of rapid immigration and emigration; many schools in these areas are the oldest in the city and are in poor condition. The textbooks too may be twenty or thirty years old, worn and unattractive. In short, the school, which is a foreign institution to these children anyway, is in addition poorly equipped, unattractive, overcrowded, and so restrictive that any hope of convincing these children and their parents that education is a Good Thing is lost in the dismal and gloomy surroundings.

The student from the ghetto has difficulty with all language functions requiring him to speak or use the formal language of the dominant culture—whether in "if-then" math problems or reading workbooks. Current textbooks for reading, which should help him to build this formal language function, do not take his differences into account, however. From the "see-Nancy-run-and-play" beginning reader to the high-school language books, most available texts are designed mainly for middle-class students who are already familiar with the dominant culture's formal language. The story in beginning readers is usually set in a suburb, which immediately makes it "foreign" to the disadvantaged child. The family is white with a mother and father, two or three children, a dog and/or cat. The children are usually blond or red-headed. They have a comfortable house with a front lawn, trees, flowers, and picket fence. They have a car, and the children have toys, wagons, and bicycles. Everyone is very happy—except the slum child reading the book. It would be neither necessary nor advisable to produce textbooks with pictures of blighted urban areas so that the slum children could feel "at home," but more books set in city neighborhoods might, for the youngest children, make reading more relevant. The vocabulary in these books could also be made more familiar.*

By junior high, most ghetto students are reading three or four grade levels below the national average. By the time they are old enough to drop out of school, thirty to forty per cent of all students do so, and the percentage of dropouts in the disadvantaged areas is much higher than this national average, usually exceeding fifty per cent.

Games in the curriculum can reduce many of the problems faced in the school by students from disadvantaged areas. They provide a mode of activity already familiar to and usually well-mastered by ghetto students. And the situations and roles played in the games can be chosen from those most salient to the students, thus maximizing their interest and involvement.

An analysis of game dynamics shows many more specific advantages of games in reducing the cultural barriers of the disadvantaged student in the normal school environment. Some of these are active learning, immediate reinforcement of learning, self-development, versatility, improved attention span, communication, and discipline.

Disadvantaged young people, as we have said, tend to react to unsatisfactory or unpleasant experiences in the classroom with withdrawal or apathy. Games supply an immediately satisfying social experience and require their active participation; students' interests and motivations are stimulated. This active response also improves the situation for the teacher, since it gives him in-

---

* Publishers frequently produce several editions of a text. The suburban and Southern editions show happy white children with a pretty teacher, the parochial-school edition shows the same children with a smiling nun, and the urban edition shows, perhaps, a Negro and Chinese student with a teacher. But these children in the city editions, though representing other races, are still basically "white" in that they do not possess strong racial characteristics beyond tan skin or slanted eyes. The atmosphere is still suburban and still foreign to the slum child.

formation on how the student is doing. Only when a student responds can a teacher accurately assess his performance.

As noted before, the cultural environment in ghettos often fosters authoritarian personality traits and poor impulse control. Playing an educational game will not change a child's personality, but games can and do stress rational decision-making, the understanding of cause-effect relationships, and the rewards of self-restraint. Moreover, through the role-playing and the simulated environment, games broaden the horizons of the children, making them aware of circumstances and relationships outside their personal experience.

Because they can be adjusted to any level of complexity, games can be tailored to the particular level of intelligence or aptitude of the group using them. The game scenario should be made relevant to the background and interests of *any* group of students; playing it will therefore seem important to them—which is often not the case with more traditional methods of instruction. As a result, children concentrate on the games and become involved with them for extended periods. This helps to overcome the problems caused by the short attention span characteristic of many children in inadequate ghetto schools.

Children who are normally shy and withdrawn in face-to-face situations, for whatever reasons, may become surprisingly active and communicative within the context of a game. This may be because all those involved are playing roles and normal relationships are suspended—risk-taking and the possibility of losing face are accepted because "it's only a game."

Lastly, a game greatly accelerates the sequence of activities being simulated. Game-playing provides an immediate reward to the individual who makes a correct decision, while the student who fails to do so knows his mistake at once and can correct his error. It therefore becomes possible to convey concepts of time perspective and the payoffs between present and future "gratifications," which are typically less appreciated among children of disadvantaged backgrounds.

On a broader level we have seen that abstract ideas can be

conveyed concretely by means of games. The mere use of symbols in setting the scenario and distributing resources to the players accustoms them to abstraction and symbolic representation.

The teacher's role in a game is not one of a dictator of class activities, but rather that of an arbiter, explainer, coach, and conceivably a player. This supportive role can assist in breaking down the negative attitudes toward the teacher common in ghetto schools. Moreover, games are self-teaching. The players learn from their own experience and that of other players within the game. It is most important that this game experience be related by a teacher or instructor to a wider framework, but to the extent that the game itself is self-playing and self-contained, it requires less teacher effort per time expended than expository methods. Even the most rebellious students seem to accept without question the rules of a game, and there is little problem with discipline.

The various educational benefits of game-playing are clearly seen in the report of Betty Rea, a teacher who observed a game of "Raid" played by fourteen students from culturally disadvantaged homes in the Boston area. These boys, participants in the Thompson Island Urban School Extension Program, ranged in age from thirteen to seventeen. Their skills in reading and writing covered a wide range from about first or second grade through early high-school level. They had experienced difficulty in adjustment and performance in the public school classroom.

The game of "Raid" consists of the interaction between a team of Police, a team of Racketeers, and (in this case) three teams representing the populations of City Blocks. The Police and Racketeer teams have resources in the form of weapons, men, and, in the case of the Police only, money. The City Block teams have men and money. The objective of the Police team is to catch the Racketeers; the Racketeers team tries to obtain money and recruit men from the City Blocks; and the City Block teams try to maintain or increase their wealth and their population. The Police and the Racketeer teams may visit (raid) the

City Block teams at certain times, but otherwise all communications must be written. The players displayed great sophistication in strategy. The game was won by the Police team.

Miss Rea offered the following observations on the conduct of the game:

> The explanation given before the game of rules and procedures held attention and checked understanding. A lecture situation was avoided by having a question-and-answer exchange. The abstraction involved in the use of symbols for people and weapons did not bother the boys.
>
> The explanation and team assignments were given in a regular classroom. The actual playing of the game took place in the school library, which allowed plenty of room for the separation of the five groups. Locating all the groups within one room conveyed the total picture of the social forces involved.
>
> The provision of seven leadership roles simultaneously was excellent. This is a situation rarely achieved in normal classroom activity, where the opportunity for only one or two leadership roles inhibits demonstration of all but the strongest leadership initiative. There were two policemen and two racketeers, as well as three captains of the three block teams. All teams included boys from both high and low achievement groups. While one of the policemen and one of the racketeers was clearly the stronger, the boys participated actively and had a chance to exert more leadership than they had shown in regular class discussions.
>
> Students of low academic achievement participated effectively, at times displaying creative thinking, leadership, and verbal expression well beyond their usual performance.
>
> Students of widely varying ability worked effectively together on teams, despite the fact that students of lower ability had previously objected to working with the higher ability group.
>
> A dramatic difference in behavior was demonstrated by one extremely withdrawn student, almost noncommunicative with teachers, by his vigorous participation in the game. This experience caused me to wonder about teacher participation in games to help reach withdrawn students. If a student could express himself freely to his teacher in a role-playing situation, it might decrease inhibitions in other teacher-pupil communication.

There was a general reaction on the part of the five teachers observing the game that the short time available for follow-up discussion (approximately ten minutes) was not adequate to clarify substantive issues. A thorough discussion period would have enabled us to judge more carefully what was learned as well as to develop further understanding based on the boys' observations. When the headmaster asked if the boys had learned anything, there was a general nodding and commenting indicating "yes." One student commented that the game taught him "how hard it is to be a policeman." Another commented that he had "never concentrated so hard" in his life. This is particularly significant because his performance in class has often been disturbing to others.

The writing of notes in playing and the evaluations written by the boys following the game allowed for a limited amount of practice in reading and writing. One interesting incident occurred when one of the block leaders was captured by the racketeers. Although the two boys left in the group were almost illiterate, they did manage to write a note to their imprisoned partner. The "police chief," having a typewriter at his disposal at his position behind the librarian's desk, voluntarily kept a record as the game proceeded.

The following quotation was taken from a student evaluation of the summer program. The student here illustrates the attitude which most hold toward classes too heavily dependent upon lectures by the teacher. "Most of the teachers I go to in the afternoon talk a lot and most of the people either don't pay any attention to what there saying or there either sleeping, where as if we had work that would keep us busy for forty-five minutes they'd get a much better responds from the kids than they do now."

There was clear evidence that "Raid" exerted a significant motivational force. The boys were stimulated and performed in both quantity and quality beyond all expectations. Their continued interest was shown by the fact that a number of the boys continued to ask to play the game again. This was especially true of some of the boys of lower achievement, including the withdrawn student mentioned earlier. Educational games can help relieve the teacher of overdependence on lectures; they have great potential for increasing student attention and involvement.

It is possible that the respect which these students feel for "the rules of the game" could be used to develop more appreciation for the unwritten rules of classroom deportment and social custom. Experience with the lower reading and language group caused me to think in this direction. This group consisted of seven boys ranging in age from thirteen to seventeen. Reading and writing ability ranged from first grade through about third or fourth; speech patterns were inarticulate and highly incoherent. On several occasions performance by these students improved markedly simply by allowing them to play a role. Role-playing and games seem to elicit a similar response; the boys appeared willing to observe in play the same rules which they flaunt in ordinary communication (for example, speech patterns improved markedly when students were depicting participants in job interviews).

Another example is the game of "Manchester," designed to be part of a high-school curriculum unit (grades 10 to 12) on the Industrial Revolution. The game deals with economic forces affecting the migration of workers from village to city during the Industrial Revolution in England at about 1800.

The game is played by seven players assuming the roles of a squire, two tenant farmers, two mill owners, two families of village laborers, and (the same) two families of laborers in the city. The players sit around a gridded game board depicting plots of land in the village, the road to the city, the city with two factories, and a poorhouse. Players have colored blocks representing their "populations" which they place on the "owned" or "rented" grid squares. All players attempt to maximize their wealth—the squire by renting out land (grid squares) and lending money; the farmers by renting land and buying labor (pieces) to place on the land squares and produce food and cloth; the laborers by working for wages on farms or in the mills (planting their pieces there); and the mill owners by purchasing machine pieces and labor and placing them in the factory squares to produce goods for a preprogramed market—fluctuations of which are unknown to the players and represented by a series of market demand cards. All the players must pay "subsistence" (play money) in each round to the banker, in proportion to their style of life. The mill owners, for example, must maintain an establishment costing several times that of the laborers. Those who cannot pay subsistence go to the poorhouse square, appropriately indicated by a sketch of a barred red-brick work-house.

All players begin with typical accumulated resources—the squire with land (squares), the laborers with modest savings (play money), and the mill owners with capital (play money) and machinery (pieces). The squire then rents land to farmers, who hire laborers to help work the land. The productivity of the land is a function of the labor invested in it, as shown by a chart. The mill owners need laborers to run their mills and must offer higher wages to the farm laborers to attract them to the city. Some of the laborers respond to the enticement and move to the city to work in the mills. Production is then increased from single-machine mills initially to several-machine mills. The mill owners, anticipating continuing high demand, tend to in-crease production until demand (shown by the demand cards uncovered) suddenly declines (historically, the impact of the Napoleonic wars and overproduction) and they must cut back. Laborers are then laid off, and those who have not saved enough money to support themselves until they can get work in farm or mill, or who don't have the "fare" to return to their village, go to the poorhouse. Because poorhouse costs are borne by mill own-ers in the form of taxes, mill owners are inhibited from following completely brutal labor policies.

This game was test-played by average high-school students in 1965. One of the designers and observers, John Blaxall, wrote:

The players were six 10th-grade students from Newton and Cam-bridge schools, plus an adult designer of the game to make up the seven. The six students were of about average ability and interests —the designer played a rather "conservative" game and did not attempt to take advantage of his prior knowledge.

The students had talked about the game . . . on the previous day, at which time the roles were assigned and the possible ac-tivities of each player discussed. They were not provided with any rules or written information about the game, although they had been exposed to some of the unit's written material (in particular the readings from Engels describing parts of Manchester, and from Bronte and Bamford describing the mills). Only the production schedules, money, and counters were handed to the players. Eleven game moves and one trial move were completed in two and a half hours. The "class" was scheduled to end after one and a half hours but at the students' insistence the game was continued—even the "final" conclusion was protested. In terms of player enjoyment, the game was highly successful.

Roles

The *Squire* was played by a rather shy person, who did not become a very active participant (though there was no sign of boredom or discontent). The Squire acted as Banker. The *Farmers'* play was "normal," though the student player only gradually became aware that maximizing output was not the same as maximizing profit. Wheat prices were sufficiently high that there was little pressure on the Farmers to cut back costs, and, owing to the use of price cards rather than demand curves, there was never any incentive to reduce production in "bad years." These matters require adjustments of fairly simple nature.

The *Mill Owners* did well (though they did not in fact "win") and were very successful in keeping city wages down to only a little above the subsistence level; in this, however, they seemed to be much more concerned with the bargaining process *per se* than with a rigorous assessment of costs and profitability. Both did become aware of the advantages of using large machines as opposed to several small ones (economies of scale) and quickly grasped the associated costs in loss of flexibility. There was no formal collusion between the Mill Owners (or between other players), though naturally each adjusted to the other's activities.

The *Laborers'* roles were played by perhaps the most dominant personalities in the game (though conceivably their behavior was effect rather than cause). They had spirited bargaining sessions with both Farmers and Mill Owners and were quick to note the relationship between jobs available and Laborers resident in the city. They were unable to drive up wages in the city, but they nevertheless seemed to make "a comfortable living," with their "families" spread about half and half between village and city. There was no continuing trend of migration from country to city beyond the early stages of the game, presumably because there was no pressure to leave the village. This was because on the one hand the Farmers were hiring more labor than was strictly profitable, and on the other hand (more importantly) there almost invariably existed the alternative of earning more than subsistence in cottage weaving. Thus there were never any Laborers—or others—sent to the Workhouse. This can be corrected without difficulty by increasing Laborers' subsistence costs or reducing the output from cottage industry.

Teacher participation was important in setting up the "trial move," but thereafter was minimal and on the level of assistance rather than demonstration. The teacher's major role was in assisting the Banker.

On one occasion, a Laborer was faced by an adamant employer in the city and remarked that it would be a good idea to form a union. It was quietly pointed out that this would be a violation of the Combination Acts, about which the students had read a little. The student's reaction was a frustrated "It's not fair!"—and when prompted by a question as to what to do to ensure work for the "family," he replied with a jesting threat to the physical safety of the Mill Owner. This set of interactions has obvious historical parallels, though some careful work would probably be needed to bring out effectively the significance for the students.

As homework the students were asked to answer briefly in writing five questions about the game. In all discussion with students it is, of course, extremely difficult to distinguish between "objective judgments" and remarks designed to respond to the teacher's or adults' expectations. It is a fact, however, that all the students were most enthusiastic about the game, thought that it had helped them to understand events, and were eager to play again.

The *Squire* felt he would like to play the same role again in order to explore the role and possible activities more thoroughly. He felt he had played wrongly but suggested that the Squire could be a "winning" role if played well—if, for example, land prices were kept high to force continued rent payments. He had changed roles and played as Mill Owner for just one move at the end of the game but obviously had not derived much understanding of "city life" from the brief experience. Asked to defend the use of games as teaching tools at the end of the discussion, he said, "Anybody can tell you something, but when you actually do it, you experience something."

The *Farmer* similarly wished to play the same role again, despite obvious difficulty during the game in handling some of the calculations. Her advice to another person about to play the role was to bargain hard with Laborers and to take the initiative in dealing with the Squire. She felt that the game helped to make clear what Laborers' conditions were like and helped players understand why they were so. She could not have afforded to pay more, she claimed—and added that she would not have done it if she could. In defense of games she said simply, "Doing's better than imagining."

One *Mill Owner* wished to play the same role again to explore further the various possibilities of the role. Her advice to a new player was to keep wages low and use bigger machines—but to be wary of running short of labor once large machines were installed. This point introduced a discussion of the extent to which the Mill Owners were faced by a labor shortage, which referred directly to readings the students had done. She admitted that a good way of ensuring a supply of labor would be to offer higher wages, but pointed out that most of the time in the game the Laborers had *accepted* low wages in the city. She noted that while the readings did suggest that wages and hiring standards were low, they did not say *why* they were low. The game "made it much more realistic," and showed that the Mill Owner was perhaps not such a heartless person. Her comments in defense of the game were "In reading, the name of the person is given to us, and we don't have time to put ourselves in his place—playing the game makes it real. . . Reading you get distracted, but the game really makes you concentrate."

The other *Mill Owner* would welcome an opportunity to play Farmer in a future game "to find out what happened at the other end of the table." She noted that one reason why wages could be kept low in the city was that Laborers had "family reserves," and this point was expanded to demonstrate that the more individuals there were acting independently, the more pressure they would feel to get employment. She found herself, she said, acting just like "a cruel Mill Owner," but realized there was no alternative. Defending the game, she commented, "Doing it, you really take the place of the actual person."

Both *Laborers* thought that they had the best roles, especially since they were involved in both the city and the village, but one wished to try another role for the experience. They were both clearly aware of the problems of being a poor Laborer, and despite the absence of continuing higher wages in the city, somehow felt that the secret of winning lay in forcing up wages there rather than in the village. One of them noticed that the Mill Owners, despite their profits, still had to borrow money, and did not seem to be overly rich. The first remarked in defending the game, "It makes learning stick better."

A revised version of "Manchester" was also played with a group of disadvantaged girls at the South End Settlement House in Boston in 1965. What was particularly interesting here was the diversity in achievement levels that apparently could be accommodated in a single game. The ages of the girls ranged from eleven to sixteen and their achievement levels ranged from functional illiteracy to college entry. The most advanced and the most backward both learned from the game, although they learned very different things. The backward student, quite without the intention of the designers, learned for the first time what interest on money was. The advanced students learned about the effect that fluctuating demand has on prices and production and some of the dynamics of urbanization during the Industrial Revolution.

The report on this test play with a disadvantaged group follows:

The game was introduced very briefly and role assignments were made on the basis of where people happened to sit around the board. No rule-sheets were handed out, but it was quickly explained to each player what he could do in the game (e.g., for Mill Owners: hire labor, buy machines and thereby produce cloth). . . .

The girls quickly lost their shyness and entered fully into the bargaining process, becoming quite deeply involved in the game.

The production decisions were rarely economically rational, but it was interesting to observe that there nevertheless developed within the game a movement of Laborers away from the village to the city—one of the historical features the game was designed to illustrate.

The game was played for about an hour and a half. By that time all the players had a grasp of what was going on. At the end of the game the players counted up their wealth and a short "debriefing" was held. The responses were not as animated as had been behavior during the game, but a few interesting points did emerge.

1. Several girls had apparently learned for the first time about the cost of borrowing money (interest) through observing the transactions between the Mill Owners and the Squire (who is endowed with a large amount of capital). They were unexpectedly intrigued by this aspect of the game, partly no doubt owing to the strong personality of one of the girls playing the role of Squire.

2. Probably for this reason, all who responded chose the Squire's role as their first choice for a hypothetical future game, except the two Farmers who wished to play that role again.

3. They liked the excitement of making and then losing money, but were sometimes confused by some of the calculation necessary.

4. Asked to say how the game was realistic, one girl suggested that it was hard to make a lot of money; as an example of its unreality, someone remarked that "In real life you have to wait longer to get a job."

5. Most of the players wanted to play the game again.

The players obviously could not identify with their roles in the same way as could high-school students familiar with the social and economic conditions of the historical period. It is also clear that without adequate preparation of the players, the version of the game used involved too many calculations in the beginning. The girls (or at least many of them) were perfectly capable of making the necessary calculations but were confused by too many new things at once. The game would be improved by starting with most of the conditions fixed and then gradually introducing the variables and the necessary calculations.

Although it is possible, then, for students to understand a difficult abstract concept from a game designed for a more sophisticated audience, this is not, of course, ideal. Yet the ideal of a specific game tailor-made for a specific group of students is not always possible in a large educational system attempting to teach a unified curriculum to a great many students of different abilities. One way to make game-playing a uniquely successful activity for students with a wide variety of abilities and backgrounds is to allow the teacher to do his own custom-tailoring. Increasing the number of variables and adjusting the subsequent game play to take this increase into account, provides a means by which the game can be made more complex. The game can then be tailored to specific age levels, ability levels, and cultural backgrounds.

A further advantage in gradually increasing the number of variables is that respect for complexity is learned. A teacher who wishes to teach a particular concept can start with the simplest form of a game concerning it, and subsequently increase its complexity until students gain a clear picture of the many forces producing certain results. In this way, the teacher helps the students to avoid a simplistic view of a complex problem.

"Neighborhood," a game designed for the Curriculum Development Center, Wellesley, Massachusetts, provides a good illustration of the possibilities of custom-tailoring. This game illustrates to fourth and fifth graders some of the basic concepts of urban growth, such as requirements of housing, transportation, shopping, and employment, and the tendency of ethnic and socio-economic groups to segregate.

The basic game is simple. A gridded board contains an outline of a cleared land area bordered by river, bay, and forest. Four teams are given pieces of cardboard of different shapes representing housing, factory, stores, and cultural facilities (churches, schools, libraries, etc.). Each team's pieces are a different color. The objective is to settle and control the most desirable land—in terms of convenient spatial relationships of homes, work, shopping, and cultural activities. The placing of pieces on the mapboard is constrained by rules representing typical desires for nearness of homes to stores and cultural facilities, and modest separation of industry and housing.

To show the effects of increasing population, the players' pieces increase arithmetically at every turn, each turn representing roughly a generation. (Geometric increase would be more realistic but expands the requirements of the game board impractically.) The student players in effect recreate the evolution of a city. At first there is plenty of land, and deployment of housing and facilities is rather arbitrary. Gradually, as the available cleared land fills up and costs are imposed on teams for clearing more, the need for more careful planning becomes apparent. Teams begin to cooperate and to bargain with one another over land use and access and may even develop zoning laws.

"Neighborhood" can be expanded to include many other variables. But even the basic game teaches elementary students concepts of urban growth such as locational requirements of housing, employment, transportation, and shopping. The more complex versions include such concepts as taxation, rights of eminent domain, provisions for public utilities and public services, zoning, and depreciation of buildings. The teacher is provided, then, with a vehicle for custom-tailoring a game to his own audience and for illustrating a variety of concepts.

V

GAMES FOR
OCCUPATIONAL CHOICE
AND TRAINING

AWARENESS and understanding of occupational alternatives is a problem regularly facing adolescents, parents, guidance counselors, industrial personnel managers, and training directors. Students today customarily make the important choice of whether to prepare for a vocation or for further education in the ninth grade, when they are only fourteen years old and have no accurate idea of how the choice will affect their lives. In addition, many students who go on to further education find that this does not prepare them to do their life work as they perceive it.

The choice of an occupation ranks in significance with the choice of a mate. Freud advised that in making these two basic choices in life, consideration should not be limited to rational analysis but should also include the deepest intuitions. Intuition tends to be more fruitful when based on broad factual knowledge and experience—and experience is precisely what most students choosing a vocation do not have. The persons concerned with job counseling and guidance, on the other hand, have little or no experience with the preferences and behaviors of the persons they counsel. (Or, all too often, with current vocational developments; how could any guidance counselor keep up with the hundreds of vocations and the many different types of opportunities in each?) Effective occupational choice and training require experiences on the part of trainees, trainers, and counselors that they do not have. To some degree, these experiences can be provided in simulated form in games.

A start has already been made by Professor James Coleman and Dr. Sarane Boocock, at Johns Hopkins University, with their "Life Careers" game. This game offers the players some ideas of their major "life-style alternatives," although it does not deal in detail with specific vocations. What are still needed are a series of simulation games that offer condensed and clarifying simulated experience of actually working in an occupational role. With such a series, any student, parent, counselor, or employer could sample various occupational games to explore quickly the nature, opportunities, requirements, and rewards of specific occupations.

For such a simulation game to be successful, the players should discover the following characteristics of the occupation, probably in the following order of interest:

1. *Nature:* type of activity (physical, mental, or mixed); whether solitary or group work; degree of autonomy; location (urban or rural or both, indoors or outdoors or both); types of problems solved (technical, interpersonal, managerial); environmental characteristics (noisy or quiet, dirty or clean, formal or informal); and colleague characteristics (socio-economic level, educational level, sex, age, interests, personality).

2. *Rewards:* pay; status; security; life style; control over events, things and people; and self-fulfillment in social, emotional, and intellectual terms.

3. *Entry Requirements* at various levels of the occupation: age, experience, education, attitude, socio-economic level, appearance, etc.

4. *Opportunities for Advancement:* probability of various types and degrees of advancement as a result of effort, achievement, social skill, and luck.

5. *Opportunities for Self-Realization:* autonomy; self-expression; self-development; learning from others and the environment; aesthetic, intellectual, and spiritual values. For persons going on to college, these values may be decisive and need to be explored in some detail, because the other four sets of characteristics hardly discriminate between, say, dentists and psychiatrists, or between accountants and engineers.

These requirements need to be translated into specific game characteristics. The players should engage in "play" activities that simulate the nature of the occupation. Since occupational activities and environments are sometimes impractical to simulate physically (as for poets, housepainters, or astronauts), the game should focus on cognitive and interpersonal aspects. Nevertheless, for a taste of the job's atmosphere, it would be possible and inexpensive to project a sound film on one or more walls of the simulation room providing the background sights and sounds of the particular occupation simulated.

Again, since the rewards of various types of work cannot be simulated directly, these must be the symbolic objectives of the game. The occupation game might be "won" by the player(s) realizing his own ideal mix of pay, status, and amount of leisure time.

The entry requirements for various occupations also cannot be represented directly in the game context. But the achievement of these entry requirements can be simulated by a sequence of educational, experiential, and other "filters" to be passed through before achieving a given occupational level.

Opportunities for advancement can be shown by branching choices being offered from time to time with a variety of reward levels for each alternative. If a player is not advancing as he desires in terms of rates of pay, responsibility, status, etc., he should be able to elect lateral or leapfrog strategies. The lateral move may be to a similar position in another organization or in a related activity where opportunities for advancement seem more favorable. The leapfrog strategy may be elected where lateral moves are impractical due to isolation and could consist of efforts to improve education or performance.

Game simulation of the opportunities for self-realization pose the greatest challenge. It is the nature of self-realization that it occurs gradually. But the fantasy identification so common in game play can help to effect this feeling, at least indirectly. A variety of statements suggesting the type of satisfactions achievable can be communicated to the player both directly and by brief, suggestive audio-visual presentations.

The over-all time context of the game must be limited to a few hours—three or four at most—for practical reasons. This period should simulate enough time to permit the player to play out major changes in his level of work, responsibilities, activities, etc., to show most of the possible range of achievement in one occupation. Ten or even twenty years of development are needed to reach "mid-career."

The spatial context of the game need not be limited to any particular location but can include very different geographic environments, as long as they all are relevant to the occupation being simulated. A chemical engineer's simulated career, for example, might include the locations of a New York office, a Houston refinery, a Saudi Arabian oil field, and a Midwest chemical plant. These physical contexts can be simulated by audio-visual displays of, for example, the background sights and sounds of the office, the refinery, the oil field, and the chemical plant. A whiff of the smell of the refinery and a touch of the heat of the Arabian oil field could even be added for realism.

The level of functional detail to be included depends on the complexities the player can absorb in the available time, and the demands of substantive presentation of the nature of the work. Some typical everyday decisions and negotiations should be included to give a feeling of the details, while a sampling of more important decisions that arise only monthly or annually should also be included. Not all of this can be done in a few hours, but a well-balanced sample of typical major and minor activities must be scheduled. If some minor activities are too time-consuming even in simulated form, their behavioral results might be presented for simulation, such as the resulting boredom, fatigue, irritation, excitement, or worry.

The roles in an occupational game depend on the number of players available and the diversity of their interests. If a single student wants to explore a particular occupation, then it will be a single-player game "against nature" or chance. If several are interested in the same kind of job, they can play the same game competitively, taking on various roles. In either case, the game would consist of a series of career decisions presented in printed or audio-visual terms, to which the student would respond by a choice, the consequences of which would then be presented to him along with further decision alternatives. If several players are involved, their interaction should be through the written or audio-visual media of the solitary game, to maintain the feeling of there being many widespread peers.

The game objectives or win criteria should probably be set by the players themselves—whatever mix of remuneration, status, security, self-realization, power, independence, and leisure they prefer. One player might strongly value monetary rewards over prestige and self-realization. Another might want maximum self-realization, leisure, and independence at the expense of money, status, and power. Some might want to have all of these things in about equal measure. When players must determine their own win criteria, they must also consider explicitly just what they want from a job—and for many, this might be the first time they have consciously analyzed their life aims. This in itself encourages a more rational choice of occupation and more realistic attitudes. The very fact that some goals are, at least in part, mutually contradictory is worth learning in an explicit occupational context.

To constitute genuine win criteria, these game objectives must be defined in measurable terms. The occupational aim of maximum monetary rewards is simple to measure, but self-realization is not. Status can be measured according to the sociologically typical degree of prestige offered (e.g., a doctor has more prestige than a garage owner earning an equal income). Power can be measured by the number of people and dollar revenues supervised or controlled in a given job. Leisure can be measured by the number of free, nonworking hours a week; independence, by the inverse of the number of supervisors above one and the rules imposed on one's activities. Security can be measured by the minimum mathematical expected value of earnings to retirement, and the probability and degree of physical incapacitation from accidents. Self-realization might be measured for some by the growth in both scope and diversity of activities supervised by oneself (as in business or politics); for others by the increase in significance and precision of comprehension (as in scientific research).

Players attempt to gain their objectives or win criteria by expending certain resources as effectively as possible. These resources consist at first primarily of personal time and energy, but later might also include resources of security, savings, and peace of mind. For game purposes, these resources can be assigned to the players at the beginning of the occupational game and expended as the players prefer; there could be time chits, security chits, and so on.

The decision criteria by which the players decide how much of their available resources to expend on which intermediate goal at what time, are, of course, just one of the things to be learned from the game. Some tested policies might be suggested to the players to save them time otherwise lost in trial and error.

"Machinist" is a single-player simulation game that allows the player to make career choices over a simulated twenty-year period. Each successive job in the various machining occupations demands increased skills and experience while offering more money and prestige. The objective is finally to attain a job that is interesting, high-paying, prestigious, and demanding of the highest skill level he can achieve.

The player begins by taking the first of a sequence of career-profile folders, each of which describes the duties and skills of a particular position. After the player completes a job, several paths are open to him; a list of new career choices, included in each folder, tells him what abilities or experience he needs for each alternative. Each profile folder includes a black puzzle piece labeled with the player's current career position. The puzzle piece has three slots identifying the possible choices for his next move. When the player makes a choice, after reading about the alternatives open to him, he attaches a colored arrow link into the appropriate slot of the puzzle piece. This arrow tells why the player made the move he did. If, for example, he chose to become an Industry Apprentice, his arrow reads "no training," indicating that this is what leads to the apprentice position. In each list of "next moves," he is told which arrows he may choose and toward what career positions they enable him to move.

At different points in the game, the player is instructed to pick a chance arrow which may help him decide which move to take next. If he uses his chance arrow to make a move, he places it in the puzzle piece. But, if he does not use the arrow, he places it aside and does not put it back in the box he took it from.

For each year of his simulated career, the player fills out a leisure time chart and a yearly pay chart. He must decide how much time he will spend attending night school, working overtime, or enjoying leisure activities, and then puts an X in the appropriate box on the chart. If, for example, it has taken him six years to become a first-gear Machine Operator Class B, working six hours of overtime a week, he would be earning $5819; he would then fill in the block opposite $5819 and six years on the bottom line of the leisure time and yearly pay charts.

The player uses the leisure-overtime trade-off chart to see what his weekly or yearly pay is at each skill level. For each pay level, he is given a set number of leisure hours each week. If he must go to night school, he deducts the time spent there.

The decision tree which results from the player's moves might resemble the following:

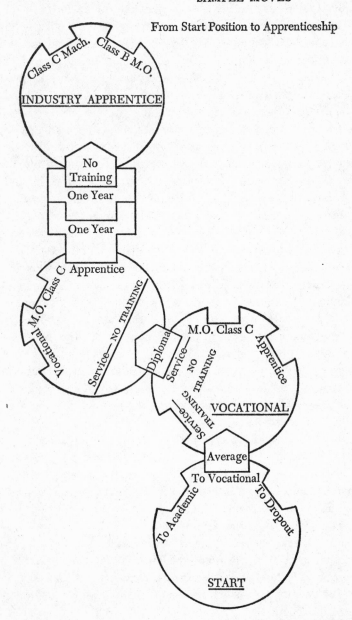

SAMPLE MOVES

From Start Position to Apprenticeship

In its first stage, this simulation was not a game but simply a flow chart to show major decision nodes in twenty years of a machinist's career. It emphasized specific decisions about the kind of work a machinist does and the specific industry he works in. The final version of the flow chart described the major decision nodes in terms of the criteria for progressing from one to another skill level. It showed, for example, what ability and experience were necessary to move from Class B to Class A Machinist or from Class A Machinist to Tool and Die Maker.

The flow chart was then turned into a game in which the student learns how his choices affect the pattern of his career.

As the game was being designed, a suggestion was made to have a flow chart covered by pieces which could be removed to reveal discrete parts of the flow chart. This would have required much repetition in the branching of the flow chart, however, since the same options could become available at different stages and from different routes. Instead, each step in the machinist's career was represented by the puzzle pieces, which represented the problems in a more coherent and efficient way.

Rather than having automatic progression from one career step to another after the acquisition of a certain number of years in a position, contingencies were introduced. Chance determined whether or not there was a job opening and whether the player could be promoted. The probability of receiving a promotion was greatest for the man of outstanding ability who worked in a nonunion shop and smallest for the man of average ability who worked in a union shop. In addition, there was a chance of a lay-off from a nonunion shop and a strike in a union shop. These chance contingencies were introduced in the form of arrow links which connected one career-position puzzle piece to another.

This game could be computerized. It is possible to envision a number of occupational programs stored in a computer bank connected over phone lines to consoles in schools and business concerns. Anyone wishing to explore a given career could, by using the console, make his decisions and receive the results quickly. The many years lost in the present trial-and-error method could then be reduced and the person unsure of his vocational goals saved much grief. Even in the area of vocational choice, then, games or simulation processes allow us to explore alternative choices and analyze their results.

VI

GAMES FOR PLANNING AND
PROBLEM-SOLVING IN
GOVERNMENT AND INDUSTRY

THE GAMES so far described have been intended to be descriptive of general and repeated situations that constitute some major issues dealt with in education and occupational training. In the government or in industry, however, one is usually interested in learning how to analyze or solve concrete and specific problems. Frequently, experimentation and lengthy, rigorous analysis are precluded because the costs would be too great. Frequently, too, the dynamic processes involved bring about such rapid changes that the problems themselves seem to defy permanent definition.

The chief advantage in using games for governmental and industrial planning and problem-solving is that they provide a mode of experimentation with alternative strategies and tactics in a competitive and constantly changing environment. The fluid nature of a game approximates the uncertainties encountered in a real situation. The controlled circumstances which the framework of a game imposes build into it certain relationships between the decisions made by the players and the results. But these relationships are unknown initially to the players and are only revealed as the game proceeds. So the outcome of a particular decision or strategy has immediate effects, thereby providing data for analyzing and evaluating the selected course of action with neither the time-lag nor the potentially irreversible consequences of a decision in the real world.

To illustrate the process or problem to be analyzed and to create a vehicle for conveying substantive information, a basic plot or conflict situation is designed. The game designer then decides how much information each player will need in order to make decisions and what substantive information to provide. This information is given to each player by means of a *profile* and *scenario*. A profile describes the character of the individual or team and states the influences motivating the player's behavior: his objectives, values, knowledge, and acquaintances. A scenario describes the over-all environment in which the game takes place.

After every game played, the game administrator helps the players to analyze their experience by leading the discussion and asking key questions. What kinds of decisions were made? What were the effects of such decisions which became immediately apparent? What constraints did teams or players feel? What influenced the decisions made? What kinds of interactions occurred between teams and players? What did the players learn? What did they feel they had done wrong? What course of action might they choose next time? Did the game fulfill its stated purpose? How closely did the game situation approximate a real one? What kinds of uncertainties were experienced? What personal reactions did players have to each other in the context of the game?

The game design depends, of course, on the objectives of the organization sponsoring the game. When the chief objective of the game is to illustrate the complex interplay of numerous variables in a given situation, so that the players have a better idea of the factors affecting any outcome, a free role-play is generally best, in which players are given profiles and scenarios and begin interacting immediately. The activity is not "programed" unless one wishes to explore the results of a particular stimulus to action, in which case the scenario describes a particular crisis or conflict situation on which the players must act freely.

The three examples of games for the government which follow are all free role-plays. "Corridor" was developed to explore the technological, economic, and political constraints on alternative plans for Northeast Corridor transportation. "Politica" was developed to explore the dynamics of prerevolutionary crises in a Latin-American setting. "Simpolis" was designed to investigate major urban issues, possible responses to crises, and some consequences of crisis decisions.

"Corridor" is a computer-assisted human-player simulation game that explores the political and economic factors which come into play in the formulation and implementation of regional transportation policy. The game decisions are made by the players, but in areas of complex calculations, such as economic consequences of moves, the computer assists in processing the data resulting from these decisions. The area simulated in the exercise is the Northeast Corridor, or megalopolis, running from Boston to Washington, D.C. For simulation purposes, the Corridor was divided into major urban centers (or "nodes"), the states they were located in, and several multistate superstructures. Actors played the roles of federal officials, state officials, representatives of the transportation industry, and representatives of industries that are transportation consumers. The federal actors were a Presidential Special Assistant with responsibility for over-all policy determination and budgetary review; the Secretary of Transportation, representing the Bureau of Public Roads and the Federal Aviation Agency; the Director of the Office of High Speed Ground Transportation, a division in the Department of Transportation; the Secretary of Housing and Urban Development, representing, for purposes of the simulation, the countervailing interest of metropolitan transportation, and including functions of H.E.W. and O.E.O. with specific urban orientations; the Secretary of Labor, who acts primarily as a mediator and spokesman for government policy in any labor disputes that may arise; and representatives of the Civil Aeronautics Board and the Interstate Commerce Commission. Each state was represented by a governor; mayor of the principal city; Senator or Congressman; a public authority responsible for revenue-producing transportation facilities, including toll roads; a state transportation executive, typically involving both public works construction and public utilities regulation; and a city or state planning agency.

The transportation industry was subdivided into its rail, air, road, and water modes, each represented by a single actor controlling the full resources of his particular mode throughout the Corridor. In addition, each mode was also represented by a labor union.

The simulation consisted of an economic submodel and a political submodel: the economic submodel concerned with the operation of the transportation industry and the intercity flow of goods within the Corridor; the political one with all other factors affecting the planning and implementation of regional transportation policy—politics in the narrow sense, public policy, labor-management relations, and the physical, as well as financial, nature of the transportation system.

Following a general briefing, short policy statements were made by all federal, state, and local officials—to give the actors a dramatic indication of each other's public positions, and to provide the occasion for some humorous, iconoclastic breaks from the lengthy briefing. Political and economic activities were then carried on simultaneously in free discussions, meetings, and lobbying, with coalitions forming and dissolving while the needs of the Corridor for a transportation plan were planned and negotiated. The ultimate objective of each player was to develop a transportation plan both suited to his own needs or intentions and closest to the ideal plan acceptable to all other influences and powers represented.

The economic submodel had the primary objective of familiarizing the industrial and carrier actors with the existing transportation system and giving them a chance to suggest how it could be improved. Game play involved the industrial and carrier actors each trying to maximize profits—industrial actors by selling inventory and using the most efficient transport mode to ship their goods to available markets, and carrier actors by persuading industrial actors to ship as much of their freight on their carrier as they could handle. Carrier actors had to operate within the constraints of limitations on capacity; industrial actors within the constraints of allowable market demands within each node and of satisfying the speed-*vs.*-cost shipping considerations which affect the sale of their products within any one market situation. The "control" or referee group acted as a generator of feedback to the industrial actors, letting them know after each period of play how profitable a choice of market and mode mix they made. (This is the computation that is made with the aid of a computer submodel of the economics of the game.)

After the period of simultaneous economic and political play, the actors divided into *ad hoc* planning groups to develop formal plans. At the same time, a federal planning group—consisting of the Special Assistant to the President, the Secretary of Transportation, and the Director of the Office of High Speed Ground Transportation—formulated a plan which was then presented at a debate on plan adoption and voted upon by all players. If the plan was ratified, this period of play ended. If not, its opponents were allowed to explain the reasons for their opposition, after which the federal planning group retired to reformulate the plan in order to meet these objections while still preserving their essential aims.

Following ratification, the constraints of the economic and political models were formally revised to reflect the plan as adopted. The players must now "live with" the plan as it is carried out. This period is, in essence, a replay of the combined economic game and free role-play under new conditions. This replay includes an opportunity for limited modification of the plan as play progresses. The winning actor is the one whose original plan comes closest to the plan which exists at the end of the game. Thus the game is a test of one's ability both to plan persuasively and to predict the responses of other parties.

One of the main objectives was to discover the likely relationships between local and federal actors in the planning process. On two occasions when the game was played, similar planning dynamics were observed. Federal actors initially ignored local issues when formulating their plans (even though some of the federal roles were actually played by local officials). Provisions for urban facilities were tacked on piecemeal to the resulting plan in order to obtain ratification. Local actors, on the other hand, tended to attack local problems and support local improvements rather than to bargain with other actors for joint regional improvements. (Again, despite the fact that some of them were federal officials. Role seems to be a more powerful determinant of action than experience when the two conflict!). In both games, the final plans reflected a lack of cross-modal planning. As a result, the implemented plans produced unbalanced, uncoordinated transportation systems which attracted more urban traffic than they could accommodate. Also, the federal actors spent an unrealistic percentage of the federal budget on the Northeast Corridor region.

The game as it now stands cannot be used to obtain information about management of a transportation system, implementation of specific plans, or similar matters. It was not intended for these purposes. But in exploring and illustrating the various political and economic factors which affect the formulation and implementation of regional transportation policy, the game was eminently successful.

"Politica" is a human-player simulation of the process of internal national conflict leading to either democratic change, revolutionary change, or reaction. It simulates the economic and political functioning of a nation by structuring the roles of major interacting national groups, placing them in conflict or cooperation in a game situation, and identifying from the result the societal and human variables relevant to a study of incipient insurgency.

The game is played by role-players acting in response to pro-gramed objectives and to various crises that develop from their interaction. The players form coalitions, negotiate with one an-other, move about the simulated country's regions (by moving from room to labeled room), and symbolically execute various ac-tions within their power—riots, bombings, arrests, curfews, broadcasts, etc. These actions are feasible or not, depending on the actor's voting, population, and physical power as given in his initial profile, his gains and losses since the start of the game, and particular circumstances as interpreted by the referee ("con-trol").

Each player is presented with a portfolio containing his per-sonal and social profile, as determined by the list of social varia-bles; his voting and population strength; his "power units"; and an economic transaction sheet indicating his last year's income and leaving a space for him to record his formal and/or illegal economic transactions.

The game is scored by awarding points (and penalties) to the actors for performing actions which further (or hinder) their over-all strategic objectives. At the same time these actions may result in negative reactions by other players and so negate the apparent strategic advantages. Points can be scaled differentially to reflect the varying degree of interest by various actors in the specific actions. The winning player is then the one with the greatest number of points.

In general, points are awarded for attempting actions in line with over-all objectives when the attempt itself represents a com-mitment to an institution. If the action is successful, further bonus points are awarded, above and beyond the economic or political benefits or losses that might result from the reactions of other players. By the same token, if the action is unsuccessful, a penalty is assessed which results in a net loss.

"Patria," the hypothetical country used in the first plays of "Politica," was composed of functional elements which can be found scattered throughout several real Latin American nations. The country was divided into four distinct regions: the capital city, the provincial city, the industrial sector, and the agricultural sector. These four regions were physically represented by separate rooms or areas. Major actors in the game were distributed among the four regions. There were three simulated political parties, each representing a focus of ideological interests: "CD," representing the conservative party; "APRA," representing the moderate liberal reform party, and "FLN," representing the extreme leftist, radical, and/or Communist party. Any of these parties could be regulated in strength by the scenario and could have as many wings or as much support among the interest groups as desired. As it happened, the first scenario studied the reaction of an incumbent CD government to mounting APRA support concomitant with threats from the CD right wing; the second scenario highlighted an APRA government's response to a threat from the far left. Only CD or APRA could be the government—meaning that whichever party it was initially controled at least one government representative in each of the four sectors. In each region, the military and each of the three parties had at least one representative (unless changed by the requirements of a specific scenario).

Numerous test plays of "Politica" by designers, political-science professors and graduate students, and government officials, revealed that the game could simulate many different political crises and many social variables affecting political stability in a given environment. Most significantly, the game led to conclusions not obvious from the scenario alone, and offered the possibility of forecasting individual and group responses to specific environmental stimuli.

The third government game is "Simpolis," developed for a "design-in" in New York City in 1967. Players dramatically encountered seven major urban problems in transportation, education, housing, civil rights, poverty, crime, and pollution—in a setting in the city of Simpolis. The aim was to communicate the essence of urban issues to the participants, including possible responses to crises and some consequences of crisis decisions.

The game was played on a rectangular grid composed of

small groups of chairs representing city blocks and streets. (The original game was set up in a large structure on Central Park Mall in New York City.) Players took the roles of key Simpolis decision-makers representing government, industry, and interested citizens from three economic classes (upper, middle, and poor) and two races (white and black) living in the city or suburbs. Some examples of the roles which players take are mayor of Simpolis, black civil rights organization leader, traffic commissioner, Citizens' Clean Air Committee Association president, and head of the organized crime syndicate.

The game proceeded as follows: after rules were explained and players were assigned roles and teams, a series of improvised one-minute speeches were given by those players taking the roles of seven city officials. Then came a series of urban-crisis bulletins, to which the players had to make immediate responses, interacting freely with one another to develop crisis responses and reactions to responses, and generating new crises directed toward achieving their objectives.

Player activities included negotiations, coalition formation, verbal and other nonviolent protests (such as blocking streets), and finally, campaigning and voting. This last was the game's climax, involving the announcement by mayoral candidates (both incumbent and opposition) of their platforms, naturally emphasizing crisis issues and the candidates' own responses to these issues. Candidates also announced intended key administrative appointments. All the players voted in the mayoral election, their votes weighted by the population they represented, and the game ended with a tally of votes and announcement of the winners. The game was followed by a discussion of the results, including an evaluation of the most realistic and unrealistic aspects of the play, new ideas generated in the course of the game, and their implications for future urban policy.

Specific crisis situations that can be introduced into this game are:

1. A rundown school in a poor neighborhood of Simpolis is burned down (arson) and rendered useless. The largely black

student population must be relocated immediately until a new school can be built. A community action group has already formed to prevent the construction of the new school on the same lot as the old one; rather, the group wants it moved so it will be closer to a nearby poor white neighborhood and the student population better balanced racially.

2. A new elementary school has been opened in the center of a new upper-middle- and middle-class housing development. The school is five blocks from an older, largely black, lower-income housing project. The children in the project are assigned to a much older school also five blocks away. A largely black group of mothers and children, insisting that their children be admitted to the new school, begin a simultaneous boycott of the old school and picketing of the new school.

3. For the third straight year the voters have defeated a bond issue which would have paid for an increase in teacher salaries. The local teachers' union today announced that a mass strike is planned by both elementary- and secondary-school teachers until they receive positive assurance that other sources of funding will be found. By the end of the week, all of Simpolis's schools will be closed. Many working mothers will have to stay home to care for their children. While the mayor fully supports the salary increases, the school board has been adamant in its refusal to consider the matter further.

Or the game can "take off" from actual events occurring just before the day of game play, to explore alternative responses to an actual crisis. The results would depend, of course, on the issues chosen and the audience the game was tailored for.

When the object of the game is solution of a known problem rather than investigation of the factors making up the problem, the game is likely to be a controlled role-play. This differs from free role-play in that the components of the problem are known and the problem itself can be structured. In the controlled role-play, the players receive profiles, but the scenarios detail a specific problem and state the specific constraints within which the players must operate. In addition, numerous other "inputs" can be given to the players. This is more formal than free role-play, far more structured, and often includes time limits within which certain objectives must be met. It is used chiefly when the objective is to develop problem-solving capability rather than identification of problem variables.

Games for industry are usually controlled role-plays. For example, the game "Merger" was designed to illustrate certain aspects of industrial mergers: how a company chooses a merger partner, what financial arrangements must be made, and how a company's organizational structure is affected. "Superb," a computer-assisted management game for supermarket executives, requires the players to make decisions about trading stamps, hiring practices, addition of soft goods, renting or buying new stores, and taking options on land.

In "Merger," ten groups of participants play at being ten different companies, five Atlanta-based and five Baltimore-based. These companies are potential merger partners. Each "company scenario" gives complete information about the company, including maps showing its marketing regions. The motive to merge in this game is the possibility of a large contract to supply a particular product to a restaurant chain. Although it is assumed that each company will find it advantageous to merge, there are objections on the part of individual members of the executive committees. The winners are the companies which work out the most mutually advantageous deals.

In order to perform the various operations required of them, players are provided with the following game inputs: an up-to-date report of negotiations with the restaurant chain; a data sheet explaining how to value companies; and a merger completion form giving guidelines for negotiation, determination of exchange rate of stock, conversion of stock to cash, and evaluation of management personnel.

In the first period, players determine what characteristics they must look for in a merger partner. In the second period, the executive committee decides what companies meet the requirements outlined during the first period. The committee asks such questions as: How does their size compare with ours? Would we expect to be the surviving company? Over-all, do they seem to fit well with us in terms of products, distribution, and philosophy? Do they appear to be a unified company with which it would be easy to negotiate? In view of their sales, earnings, and growth rate, approximately what multiplier should be applied to them and what is the over-all value of their company? In the third period, members of the committee meet with prospective merger partners to evaluate them. If they can come to terms with a company, they then work out the organization and financial arrangements of the new company.

In "Superb," seven sales managers of a supermarket chain called Superb compete for increased sales in a market place (managed by the computer) in a mythical town called Greenville.

Each of the seven players receives an information package at the start of the game. Included in this package are: game sequence and rules describing activities for eight monthly periods and policy choices available; social and economic description of Greenville, including a map of the town, supermarket selling climate, competition, etc.; player profiles (a description of each player's personality and role in the steering committee); financial statements (profit and loss statements and balance sheets); reports on supermarkets in Greenville, including operations analysis of the three local chains: Superb, Q.E., and R & R; steering committee operation, responsibilities, and members; history of trading-stamp use by the three chains in Greenville; and decision sheets with copies for each month and with one filled in as a sample.

Each month the executives discuss policy and reach decisions. The president records the final policy decisions on a decision record sheet and turns the sheet into Control. Control enters the decisions on an on-line computer via teletype and receives financial statements in return. The financial statements, market reports, and news flashes are then returned to the players at the start of the next period (month). Scoring is based upon points awarded for monthly net profit, as well as upon bonuses for good, well-timed decisions. Points are deducted for bad investments, excessive borrowing, or nonpayment of dividends at the end of the game.

Sometimes an industrial organization will want to train individuals in specific behavior patterns. A training game designed for this purpose has both to communicate information and simulate actual experience. Through the simulation, players have the opportunity to exercise their new information and skills. The training process itself is accelerated since the costly "on-the-job" practice, which often occurs after a lecture-type training period, is reduced substantially. These training games are the most highly structured of the three types discussed here.

The purpose of the game "Bankloan," for example, is to make, for the benefit of bank management trainees, a dramatic demonstration of the importance of certain principles in making bank loans: (1) in evaluating loan applications, quantitative financial record data must be balanced against qualitative data on management; (2) interview information must be used to complement written data: and (3) to maximize over-all returns to the bank, estimated risks must be neither ignored nor taken as sufficient ground for refusing a loan, but rather should be made the basis on which to secure restrictive covenants in the loan agreement that correspond in magnitude to the degree of risk being taken.

Players take the roles of representatives of three banks seeking to make loans and three companies seeking to secure them. With the company players, the object is to obtain the needed loan on the most favorable terms possible; in the case of the bank players, the object is to maximize bank profits through deposits and loan interest. In the course of the game, each bank representative receives an annual report and considers financial data about each company, interviews its representative, and evaluates the amount of risk attached to making a loan. On the basis of this evaluation of risk, he decides on the application of restrictive covenants in order to minimize risk without over-restricting the loan and thus jeopardizing chances of his bank's getting the loan.

Another example is "Supra," designed to train six players (or any multiple of six) in scientific purchasing. The six roles are three supermarket buyers for "Supra Markets" and three salesmen from three different food manufacturers, each selling three products. During the game, the salesmen make sales calls on the various buyers. In a manner similar to that in other games described above, the players are all given comprehensive packets of information about the roles they play, the rules of the game, and the environment their simulated game is supposed to represent, as well as mock record sheets, sales and promotion material, order forms, etc. In the various game periods, the salesmen move from buyer to buyer offering their products and taking orders, which they give to Control. Control records the salesmen's commissions and returns the sales record sheets. Control then gives financial inventory sheets to the buyers, who also get inter-office memoranda from the company president (represented by Control) congratulating them on having made a wise purchase decision or urging them to do better. The winners are the salesman who has made the largest commission and the buyer who has made the greatest profit.

Games for planning and problem-solving in government and industry can, then, enable participants to analyze or practice solving very concrete and specific problems. And they give the players a chance to investigate variables in very complex organizational problems. These games can be free role-plays or controlled role-plays, with practice in specific problem-solving or planning in a cost-free environment; or training games, simulating the communication of information and decisions in the actual situation in which they must be used.

VII

HOW TO THINK
WITH GAMES
BY DESIGNING THEM

————————————————

MANY, IF NOT MOST, of the world's problems involve competition for scarce resources. The lowest common denominator of most wars, civil-rights problems, and internal political difficulties is the competition for something in limited supply. This competition has in common with games a partial conflict among actors and objectives, the identities of several distinct actors or groups of actors, and the use by these actors of their scarce resources to achieve their objectives.

But games are also useful problem-solving tools for less universal problems—problems of importance to one or several individuals but not necessarily to anyone else, problems that also often involve competition for scarce resources and are often gameable, either explicitly or mentally. The sequence of steps in game design is similar enough to the sequence of ordinary problem-solving or systems analysis to permit the game-design process itself to be used for problem-solving analysis. The advantage of thinking in terms of games is that it often leads to more imaginative and less emotional decisions.

Consider a problem in domestic relations, such as deciding whether to take a summer vacation in the mountains or at the seashore. Wife and daughter would like to go to the seashore, husband and son to the mountains. This situation can easily be gamed, the game objective being the satisfaction of one's own vacation objective, the limited resources being the good will of the other actors and the time and intellectual resources one can use in persuasion, and the win criteria being the lowest cost achievement of one's objectives. If either husband or wife insists on having his or her way, he or she may get it, but the possible cost of either having the mate take a separate vacation or go along grudgingly and sullenly would reduce the "satisfaction payoff." On the other hand, simply to acquiesce to the other mate would also reduce the satisfaction payoff. The optimal solution is obviously to persuade the other person at minimum effort to go along with one's objectives happily. The rules would obviously exclude cruel, unusual, or bizarre means of persuasion, such as the wife refusing to cook or the husband propagandizing both children in his wife's absence. The tactics involved in this game can be worked out with varying degrees of detail if detail is desired; it is probably necessary actually to stage the game to effect the detailed arguments and counterarguments.

In this homely conflict, as in many other conflicts, the following general sequence of problem analysis or game design is a useful tool. First the general context of the situation is determined. Although ultimately everything is related to everything else, a decision has to be made about the most important limits of the problem—a decision that will be influenced by time-and-effort constraints and the need for detail in the problem analysis.

The scope and detail of the problem can usually be defined in terms of geography, time duration, actors concerned, and types of actions or functions involved. In the domestic conflict over vacation locales, the scope is limited to the two alternative locations. Detail consists of the advantages of each of these, and time duration is the length of the next vacation. As can be quickly seen, this limitation of the scope of the problem may in some cases preclude compromise solutions—such as a mountain area with a large lake with a beach, or something entirely different such as a trip to Paris, which overwhelms everyone's preferences.

The next step is to determine the actors who are the principal adversaries in the problem. In the example these are chiefly the parents, but the children may also be significant actors if their shifting alliances with the parents are believed to influence the outcome significantly. If there is a dog in the family who does not "decide" but who limits alternatives, he too is an actor albeit a fixed one. In larger-scale problems, actors may be groups, companies, entire nations.

The chief characteristics of an actor are that he makes, for purposes of the game analysis, at least relatively monolithic or homogeneous internal decisions and is distinguishable in preferences and resources from other decision-makers. Internal decision processes are, for purposes of that level of game analysis, uncontested and clearly distinguishable from those of other actors in terms of objectives and capabilities and resources.

Naturally, every actor has other actors within himself. Not only do nations have contending factions, and companies differing executives, but even individuals have internally conflicting viewpoints and "voices," such as the "superego" of conscience, the pragmatic "ego," and the pleasure-seeking "id." These internal differences can be analyzed only by setting up a new, more detailed game about the conflict internal to that actor—something which sometimes leads to multilevel games in which there are conflicts within conflicts.

Identification of the actors' objectives must be done in the context of the game scope. Those objectives that are relevant to the field of play or functions represented in the game are the ones that are significant. Thus, in the example above, the main objectives that are relevant to the family members are vacation enjoyment, family unity, effects on future family relations, and perhaps vacation costs. Possibly tangential objectives might be the education and training of the children, which may be differentially affected by the choice of the vacation spot.

After the actors' objectives are identified, the resources that they may use to gain their objectives must be determined. These resources can almost always be defined in terms of time, energy, past-generated resources such as money or good will, and future bargaining power which might be strengthened or weakened as a result of feelings left after this particular game.

In gaming, it is very important not to exclude psychological resources just because they are difficult to measure. In bargaining games where bluff is involved, the "nerve" or risk-taking capacity of a player may be decisive. This can be quantitatively measured in terms of the relative degree of willingness with which players take long shots, risking high stakes for high returns on a low-probability basis. Some players are, of course, capable of taking very long shots out of sheer ignorance. Sometimes, when the risk is very great and difficult to credit, a viable strategy is even to appear to be unaware of these risks, "playing dumb," and being impervious to counterthreats by giving the impression of not being able to understand.

Once the scope of the game analysis has been determined, together with the identification of the actors and their objectives and resources, the win criteria can be determined. The win criteria are usually expressed simply in terms of the achievement of a given set of objectives at minimum expenditure of resources, or the achievement of the maximum degree of objectives within a budget of resources expenditures. Thus, in the previous example, the win criterion for the husband might be the mountain vacation, regardless of the cost of his wife's subsequent feelings, or the best compromise vacation achievable within a certain limited degree of ill will as a result of the negotiation. Conversely, the wife's win criterion might be the seashore vacation only if her husband does not object strenuously, or a compromise vacation which will satisfy both of them.

The question, "What are his win criteria?" about an adversary in a gamelike problem is one of the most important questions one can ask. It forces one to analyze an adversary's objectives and determine the degree of competition or harmony with one's own objectives. If one wants to persuade an adversary in a business, domestic, or personal situation to accept one's own point of view, one must empathize with that adversary sufficiently to identify the decisive win criteria and to appeal to them in decisive argument or other forms of persuasion. In the example, the husband's identification of the wife's win criterion would quickly show him that he could get his own way if he chose to press the point.

The importance of empathy in conflict resolution has been recognized by other writers, most notably Anatol Rapoport, who suggests that the best way to resolve an argument is for each adversary to make the clearest and most persuasive possible statement of his adversary's position.* This is actually a simulation approach to enforcing empathy in the arguer. It can be accomplished perhaps somewhat less dramatically by going through the adversary's objectives and possible win criteria in one's head.

Once the game's win criteria, scope, actors, and their resources and objectives have been identified, the dynamic sequence of possible interactions of the game can be determined. This is how the game is actually played. We know the field in which the actions take place, the position from which the actors begin (initial available resources) and the objectives that they seek (terminal positions on the field), and we can identify a variety of tasks to get each actor from the starting to the ending position. In spatial terms, we can imagine two people at opposite ends of a culvert, each wanting to go through the culvert to the other end first, and there being room enough for only one person to pass through. Each player's objective is the starting position of the other player, and the field is restricted to the linear distance between them. One can readily generate the alternative possible interaction sequences. Both players may immediately

*Anatol Rapaport, *Fights, Games, and Debates*. Ann Arbor: University of Michigan Press, 1960.

start and bump into each other in the middle of the culvert, each may wait for the other to begin, or each may try to persuade the other not to start and to wait for him to go through the culvert first. Various sequences of these activities may take place.

Most interactive problems in games of any interest are much more complex, of course, and involve many paths between initial and objective positions. Usually there are more than two actors not only in competition but also in some degree of cooperation. The easiest way to analyze the problem in terms of this game interaction sequence is by imagining the initial situation or starting scenario, the variety of actor-desired terminal situations or "win" situations, and the most undesired or "lose" situations, and then all the ways the goal can be reached from the starting position (and vice versa). The simplest way to determine this is to limit severely the number of moves or amount of time to get from start to goal; this will minimize the amount of interaction possible. The complexity of the interaction sequence is a geometric function of the number of actors and the number of sequential moves they are permitted. One can readily determine the minimum number of moves available by knowing the maximum game-time available and dividing it by the average time per move.

Once the interaction sequence has been determined, we can then determine whether it is too broad or not sufficiently rich in alternatives. If it is too broad, additional constraints or "rules" have to be added to the game analysis to further limit the scope of possible actions and the number of possible interactions. If, on the other hand, the initial definition of actors, objectives, and scope of the game have been too narrow to permit a rich exercising of alternatives, it will immediately be apparent in the interaction sequence analysis. In that case, we should go back to the definition of the scope of the game and identification of actors and their objectives and resources and perhaps increase the scope, the number of actors, or the complexity of objectives and resources available to the actors.

Following this evaluation of the constraints or stimulants to game interaction (the rules of the game), the game's physical format can be either imagined, if one wishes merely to analyze a problem, or carried through, if the complexity of the interactions is beyond the imaginative capacities of the problem solver. It is very difficult for the ordinary human mind to keep more than a few variables in mind at once, and even a problem in which there are three actors with a few objectives each, and enough time to interact and counteract each other's moves several times, runs the number of alternative interaction sequences into the thousands. For simpler problems, diagrams are often a great help. Just identifying the actors by various boxes on opposite sides of an ordinary piece of paper, and drawing arrows labeled for different types of interactions and counteractions between these actors, can greatly clarify situations. For more complex problems, actual play of the game is often necessary.

The design of games for solving simple problems, such as where to take a vacation, obviously is not essential for a person who has good training in analytical thinking and who can empathize with an opponent enough to figure out psychological objectives and motivations. But game design and analysis is a good technique for the person who often finds himself embroiled in conflicts which he cannot seem to solve, not for neurotic reasons,* but because he cannot clarify the conflict in his own mind. For this person who cannot apply or does not know the abstract techniques of problem-solving or systems analysis, game design is an easily visualized method of reducing problems to easily managed and manipulable components.

* We are aware that a person with neurotic conflicts may be the last person to recognize that he has a psychological problem rather than a logical problem. This person is unlikely to resolve his conflicts with games, but will probably merely elaborate them.

VIII

HOW TO EVALUATE
THE COST-EFFECTIVENESS
OF GAMES

A QUESTION frequently raised by skeptics concerning games is whether the game is really "worth all the trouble." Hasn't the game taken more time to teach certain concepts than other methods would have? "If Johnny X would just sit down and spend fifteen minutes learning the causes and results of such-and-such event, then we wouldn't have to bother with all these fancy educational techniques. Why, when I was a boy . . ."

The chief problem in convincing this skeptic lies in the various definitions of "learning." The skeptic often equates learning with the ability to reproduce a certain set of facts in a given amount of time. This is often equivalent to rote memorization, reproduced on a test never to be considered again. If this is "learning," it is learning of short duration and little relevance to solving real problems. The only problem solved is that of finding the minimum effort needed to respond to a test.

This book has discussed the improvements in learning possible through the design and use of serious games—improvements in motivation, decision-making, strategy, negotiation, understanding of parallel processes, planning, and evaluation, in addition to factual knowledge. Games are not and should not be regarded as replacements for individual study. But assessing the value of games and whether they are, in fact, "worth the trouble" means that we must assess their cost-effectiveness, their efficiency in comparison to other instructional and research methods.

The costs of a serious simulation game are the time, energy, and intellectual resources needed by the designer and the players. If a game has been professionally designed, these costs are figured into the monetary cost of the game. If the game has been designed by the teacher and/or players, the financial cost is mainly the cost of salaries and training of additional teachers, and the modest cost of any supplementary materials needed.

The nonmonetary costs in terms of time, energy, and intellectual resources expended by the players must be weighed against the costs of achieving the same results by other methods. It may take a teacher two hours of lecture time to convey certain information, the students two hours to study and comprehend it, the teacher one hour to devise a test on it, the students one hour to take the test, and the teacher two or more hours to grade the tests. If a game, however, can provide for the giving of information, utilization and application of the information in a problem situation, analysis of the information about the game itself, and evaluation of the players in their roles in a two- or three- or even four-hour period, then obviously the costs of achieving the desired learning through this method are less than through the conventional method.

Yet despite the fact that the costs may be less, the effectiveness must be at least equal to that obtainable with conventional methods. The conventional criterion for evaluating this effectiveness is whether the game satisfies the predetermined objectives of a given learning situation. If the objective is to give the students a certain empathy for the decision-makers in a given historical or economic situation, then a game must provide for a maximum of realism and reproduction of historical reality. If, on the other hand, the objective is to generate innovative solutions to a specific problem, then a game's productivity in generating new solutions must be evaluated and compared with conventional methods for effectiveness.

The effectiveness of an instructional game in improving the decision-making ability of students may be quite impractical to test because of the very long-term consequences involved. For example, even if a student playing an economics game does not memorize any terms or theories, or does not demonstrate any measurable change on conventional exams, perhaps ten years later he will make a much better economic decision.* To measure this kind of nonacademic improvement in decision-making, one could use more problem-solving simulations, at the risk of being accused of using games to test game learning and thus assuring a positive result. However, it might be reasonable to assume that if there is evidence of problem-solving transfer from one game situation to another, there is at least a greater probability of it being transferred to real-life problem games.

Some useful criteria for evaluating both the costs and the effectiveness of games would be the active involvement and stimulation of all players; sufficient realism to convey the essential truths of the process being simulated; clarity of consequences and their causes in both the rules and the action; "playability" in terms of the kinds of materials, space, and time required to achieve these results; and the repeatability and reliability of the entire process.

The key to involving all the players lies in the degree of drama inherent in the game itself. Ideally, the drama is heightened by the players' suspense about the outcome and their curiosity about the various interactions designed into the game. The excitement generated by these two factors in the course of game play contributes to further suspense and greater interaction as

* I am indebted to Professor James S. Coleman for this observation.

more and more players are stimulated first by the game design and then by each other.

One of the important variables contributing to the self-perpetuating excitement of a good serious game is the degree to which players can identify with their roles and objectives within the game. Character interest must be built into these roles in such a way that the very introverted player will express *himself* more freely and the very extroverted player will express more of the *character's* nature. Overly structured roles often inhibit the introvert; overly "free" roles often fail to constrain the extrovert sufficiently. The dynamics of the game and the player alliances and conflicts stimulated in it are inextricably bound to the player's identification with his role.

This player identification with his role can be manipulated in numerous ways, depending on the objectives of the game and what is known about the players. In cases where the game and role objectives parallel the career objectives of the players, the game has built-in player identification. One can then experiment with the personalities or motivations incorporated into the game roles to give the players a wider understanding of the interactions possible in a given situation. In other cases, where game and role objectives are so foreign to the players that they cannot easily identify with them, the personalities and motivations incorporated into the game roles must be made more like those of the players.

The game designer, then, has a choice of the types of player identification he can include in the game—identification with game and role objectives, identification with role personalities and motivations, or both. He is likely to choose the first when the objective is to analyze interactions and their effects on a problem situation; the second, when the objective is to expand players' consciousness of how and why decisions are made in the present or were made in the past; and the third, when the objective is to allow players to experiment with alternative decisions or strategies.

One test of a game is how much player identification does occur and to what degree suspense and excitement are generated by the players' interactions. The more excitement and more interaction there is, the more active the learning environment and the more heightened the players' awareness of the factors involved in the real situation being simulated.

Games should possess enough realism to convey the essential dynamics of the process being simulated. But evaluating the "realism" of the simulation and the degree to which it may have been sacrificed for playability is one of the most difficult aspects of game analysis. To estimate this realism requires that one know the "real" situation being simulated. Yet "knowledge" of the "real" situation is often as much a matter of individual point of view as it is of objectivity. One need only listen to experts debate about international relations or economic policy to recognize how much disagreement will arise in defining "reality." When these differences concerning the real world are brought to bear on the analysis of reality in a simulated world, objective assessment is difficult indeed. Some trade-offs will have been made to take into account the constraints of time, players, and their skills. Still, evaluation of the importance of these trade-offs will often be personal and idiosyncratic.

Designers of games have occasionally expressed concern that "professional standards" be maintained by games to uphold the professional honor of this growing subdiscipline. Perhaps they mean that only "qualified" persons should design simulations and games. That may be, but one might just as well insist that only qualified persons write poetry, drama, and novels. I personally have confidence in the ability of games players, particularly those absolutely tough-minded subteenagers, to reject superficial, dull, and spurious games. Furthermore, games are even more effective as a learning mode when designing them than when playing them, so we should not deny even the least qualified student this learning experience because of fanciful ideas about expertise. Let gaming be pervasive, let it flourish in good forms and ill, and the best will emerge. This is no art of the experts, but a universal language common to all cultures, ages, and conditions.

Another evaluative agony often raised by certain enthused but moralistic schoolteachers is that if games teach so effectively, then all sorts of safeguards must be imposed to prevent them from teaching evil or incorrect things effectively. They ask, "How can we be sure the games won't teach the wrong things?" Well, you can't be sure, any more than you can be sure that a book or a lecture doesn't teach the wrong thing. Taken to its logical extreme, this view leads to a preference for the very *least* effective teaching methods, since these offer the least threat of corrupting the young.

In general a game's realism can be assessed in terms of the degree to which it reproduces the interaction of choices involved in the simulated process. This does not mean that in a realistic simulation the outcomes are identical to those in history (or whatever other process is being simulated). On the contrary, the outcome of the game may be quite different from the real situation because the game's chief purpose has been the realistic simulation of the interacting forces. To show *all* the interactions involved, that tenuous balance of forces which leads to a particular conclusion in real life may have been distorted. Other games "cook" the results to yield an outcome identical with that in reality, but often by sacrificing the realistic interactions.

Obviously, the purpose the game was designed for and the objectives of the game used affect the game's degree and types of realism. Just as a reporter can "slant" a newspaper story by his selection of which facts to include and in which order to present them, so, too, the game designer can affect the view of reality in his game. Games are ordered dramas, and some may be tendentious. But no instructional mode can be made proof against false data.

No serious game can be successful if the players do not understand its rules, their objectives in the game, the consequences of their actions, and the reasons for these consequences. In this sense serious games *should* differ from more conventional games.

They should respond more to the conscious decisions of the players than to an outside element of chance. Though the well-designed serious game can be programed for certain interactions or results, the players nevertheless should see that it was a specific decision which initiated a set of consequences or that it was a particular type of uncertainty. Thus, "Monopoly" is not an ideal instructional game, because winning depends more on the rolls of the dice than on the decisions made.

Players often become so involved in their specific roles and in the game play itself that they do not consciously concern themselves with the wider, more universal consequences of a set of decisions. For this reason it is useful to have a postgame discussion to analyze and evaluate the forces which interacted, the over-all consequences of the game play, and the specific decisions which led to the consequences. In games where the outcome differs from historical reality, the reasons for the differences can then be considered; in games "rigged" to produce a particular result, the reasons for the rigging and the eliminated factors can be analyzed. The scope of learning, then, extends beyond the game itself and includes the analysis of the game, its components, and the context for which it was designed.

The "playability" of a game, assessed in terms of the space, time, and materials it requires, is another crucial factor in evaluating its cost-effectiveness. Any game requiring more space, time, or materials than are available obviously cannot be played, and will certainly fail to accomplish the objectives of both the designer and the user.

The degree to which the designer has brought about an economy of means in the game structure is a key to assessing the seriousness of the game. The best indicator here is in the rules and player instructions. If these are too lengthy or are unclear, they are reflecting uncertainty and lack of clarity in the game design itself. If the game requires too many players to interact, acquire information, and make decisions which, in turn, affect the interactions and decisions of other players, then the game has not been adequately test-played. If so many actions and interactions occur that the players cannot ascertain the reasons for certain consequences in the game play, then extraneous and time-consuming decisions have not been eliminated from the game structure and it is more complex than is necessary (or not enough time has been allocated to play it). The players' contact with the detailed actions of a game must make him better able to recognize the over-all dynamic processes of the reality being simulated.

Superfluous and fancy materials, pieces, or parts to a game also indicate design obsessions, where the designer has been more interested in the physical parts than in the conceptual whole. Conversely, a great mass of preparatory reading, in absence of visual aids or charts, also indicates lack of clarity and decisive logic in the game design.

Assessing the important criterion of repeatability in a game entails a consideration of the purposes for which the game was designed, the number of variables involved, and the amount of interaction and role-playing required. Effective games are repeatable with players in either the same or different roles. Repeatability is here meant in reference to the avoidance of a few soon-anticipated outcomes, not to repeating the same set of actions (i.e., repeatability is the opposite of experimental replicability). When the primary purpose of a game is to illustrate the interaction of forces in a given situation, rather than a specific conclusion to it, any number of different conclusions may arise, depending on the timing of the interactions and the points of view of the players. In particularly complex games of this nature, alternative strategies of players can be tested in succeeding plays of the game, and the differing consequences of these strategies can be analyzed and explored. The greater the amount of interaction, variability, role-playing, and therefore, uncertainty of outcome involved in the game, the more eminently repeatable it is without exhausting the possibilities.

Those games designed primarily to illustrate a particular conclusion are, in many cases, repeatable when they depend more on the players' identification with roles than with the inherent sense of suspense over outcome.

After evaluating all these aspects of the game design and play, one may then conclude by asking some simple questions: Did the game accomplish the purposes it was designed for? Did the players end the game with a heightened awareness of factors involved in the real situation that was simulated? Did learning take place more rapidly than with other methods? Did the players become aware of the simultaneous interactions of forces? Did they use and apply the information generated by the game? Was the learning material employed so as to increase retention over time?

If the answer to these questions is "yes," we still cannot necessarily conclude that the game in question is a more cost-effective instructional or research method than alternatives. Controlled experiments would have to be conducted in which a control group not playing games and the game-playing group would be pre- and post-tested. This is costly, since for statistical validity very large groups would be needed to assure essential identity of the two samples in all respects except the gaming. Furthermore, additional controls would have to be worked out to avoid contamination by the "Hawthorne effect"—the tendency of people to respond positively to anything new just because it is new, without its being necessarily better.

The answers to those questions are easy to obtain but extraordinarily difficult to evaluate. It is probably both more practical and more honest to admit that scientific evaluation of games' effectiveness will continue to elude us until we develop a quantitative theory of instruction or obtain very much more experimental data than we now have. Until then—and "then" may be a decade or two away—we should probably be content with a more literary-dramatic evaluation of games. After all, critics can tell the difference between good and bad books and plays. Can't they?

IX

THE FUTURE
OF SERIOUS GAMES

THAT SOCIETY is becoming more and more complex as it becomes more highly technological and that education for life roles is becoming increasingly important are truisms today. The chasm between the "educated" and the "uneducated" is growing wider, and the numbers of both groups are increasing. Yet fewer jobs for uneducated or unskilled workers exist now than in the past, leaving a higher percentage of poor people out of work. In addition, the originally elitist concept of "individual worth" is becoming such a pervasive slogan in all areas of society that people are now demanding more attention for their problems and becoming impatient with educational and social organizations for failing to respond rapidly enough.

We will continue to have these problems in the future. Serious games will not "solve" them. But serious games are both feasible and desirable in alleviating many of the organizational, structural, and administrative problems which so often delay action on individual or specific group problems. And they are also useful in illustrating the interactions of forces and in analyzing them in an emotion-free environment. They are, then, both instructional and illustrative, valuable to the planner and to the "planned for."

One of the keys to avoiding greater social problems in the future lies with the ability of the schools to handle their increasing burdens. A student who drops out of school because it does not seem relevant to his life, because he does not understand the material being taught, or because school forces him into a passive role, will look elsewhere for relevance and action. He is not likely to find a satisfying job, and he is likely to blame not only the schools but "the outside world" in general. Ironically, he will be totally dependent on this "outside world," which will further increase his resentment.

We have talked about the ability of games and simulations to improve motivation and to relate the learning environment more specifically to the real world. The growing trend toward increased game use in the classroom is likely to continue into the future as schools seek additional ways to make learning active, relevant, and exciting for students and teachers and to lower the barriers which often make school "foreign" to young students.

Increasing attention is being given to the matter of making education more effective in urban ghettos and rural slums. As welfare costs and police costs surpass education costs in most of our major cities, the point will be brought home to middle-class voters that absolutely nothing has been saved by economizing on education, and that much—even lives—have been lost. Increasing attention will be given to effective education for impoverished minorities such as Negroes, Puerto Ricans, Indians, and Mexican-Americans as the only viable alternative to violent social conflict. To make education effective to the disadvantaged, we need active, relevant, highly motivating, efficient instructional techniques. If, as this book argues, simulation gaming is such a technique, its use is likely to be enormously expanded.

It is not difficult to imagine a school of the future as a "labo-

ratory school"—a school making massive use of educational simulation games, laboratory activities, and creative projects—a
school in which almost everything to be learned is manipulated,
physically or mentally. Students will have the chance to investigate their subject matter, to feel comfortable with it, to familiarize themselves with it, and to do so in communication with other
students, thereby giving to all students the benefit of additional
ideas and insights. The students will be in an almost totally active learning environment, exploring and discovering for themselves.

The extensive use of games and gamelike activities in the
classroom will not obviate the need for teachers. Basic knowledge must still be imparted, adult wisdom and perception must
still be exercised, and special help with both academic and personal problems must still be available in abundance. The teacher
must decide in what order concepts can be taught most effectively, by what method they can be communicated most memora-
bly, and at what point review and evaluation are needed for
"closure."

The demand for quantity and quality of teachers is, however,
already outrunning the supply. Perhaps the greater specialization
of function possible through increased use of games in the classroom will attract to the teaching profession many highly qualified individuals who have previously been discouraged by the
numerous mechanical, nonteaching activities that take up such a
large percentage of teachers' time today.

If the teacher shortage continues as the population rises, class
sizes will undoubtedly grow and teacher attention to individual
problems will decrease. To solve this problem in the conventional classroom, only two major alternatives will exist: mechani-
zation of certain types of instruction by computer and other electronic techniques; and delegation of many instructional tasks to
students.

The current economic limitations on large-scale computerized
instruction probably preclude this method from solving the problem alone in the near future. In addition, the intrinsic limitations

on any form of instruction which lacks human warmth, understanding, and sensitive responsiveness may permanently preclude its large-scale use. The quick student gets many positive responses from the computer in a given length of time; but those responses will never be as diverse and imaginative as a good teacher could give. The slower student, who needs encouragement and praise, gets far fewer positive responses in the same length of time from a computer, and also misses the more sensitive response to his errors which only a human being can offer.

The second alternative is more promising and much cheaper. And indeed, this type of program has nearly always been successful. In the Homework Helper program in New York City, for example, which allowed high-school students to tutor elementary-school students, not only did the elementary students' performances improve but that of the high-school tutors improved even more. Many other examples suggest that one of the best ways to learn is to teach and that students learn best from their peers.* If this is the case, as it seems to be, we can expect that students will increasingly be teachers in the classroom.

Peer teaching and near-peer teaching (older students teaching younger students) will demand many interactive techniques that turn the classroom into more of a laboratory, with many small problem-solving groups. The combination of reduced authoritarianism in the schools as a result of increased student instruction, and the urgent need for increased efficiency of instruction as a result of the shortage of highly specialized teachers, will probably encourage the further use of games or gamelike activities in the classroom.

Games, then, will probably be used in the classroom with greater frequency for either or both of two future developments: Even where there is no marked shortage of teachers, games will be used because they make learning more efficient, active, and relevant. In the event that the teacher shortage increases, games will be used because they provide more opportunity for student

* See, for example, James S. Coleman, *The Adolescent Society* (New York: Free Press, 1961).

interaction and peer-teaching, and because they increase the time the teacher has for individual help and counseling.

As education is seen more and more as the key to reducing problems of unemployment, poverty, and welfare, and as social, religious, and industrial organizations continue to turn their attention to adult education and training for the disadvantaged, we may expect a dramatic increase in the number of educational programs available for adult dropouts. Already the number of people who wish to enroll in these programs far exceeds the number of opportunities available. Often the programs available are open only to employees of the sponsoring corporation, and often, too, a certain basic skill level is required for admission. As efficient educational techniques continue to be developed and more funds become available for these programs on a large-scale basis, we may expect to see the incorporation of many games and simulations into the programs.

Most of the adults for whom the programs have been designed were unsuccessful in the traditional school environment. Because their desire for upward mobility and their motivation are so strong, these adults are frequently successful in adult education and training programs despite the repeated use of these same obsolete techniques. Historically, however, retention of "learned" material is less with the lecture method than with those methods by which students interact with others and actually use their newly developed skills in a realistic or real environment. Games, of course, provide an environment in which students can try out their skills without cost in the real world and in which they can see the interactions of many forces or concepts. Because they have the opportunity to apply and test their knowledge, the students' retention increases.

In addition, as these programs expand, we may expect a tremendous demand for highly specialized teachers. For the same reasons that public schools are likely to resort to increased game use to counteract this problem, adult education and training programs will probably also use games and simulations with increasing frequency.

Students, parents, and teachers are demanding with greater urgency that educational curricula better prepare students for the social, economic, political, and technological roles they will fill in the real world. The current widespread rebellions of students in both the high schools and colleges, parental concern with education on all levels, and growing interest in adult education all indicate the public feeling that education must respond to the changing trends. The public is becoming far more interested in the management of education than ever before, and it is demanding a greater role in this management.

Moreover, nearly every school system recognizes the existence of segregation, *de facto* or otherwise, and nearly every system is under tremendous pressure from concerned citizens, parents, and the federal government to do something about it. Yet the question of what to do is not only a question of morality. It is also a question of cost: every school official worries about his budget, and many also worry about re-election. Every parent worries about the effects of actual or projected changes on his child. Change is slow—schools do not become racially balanced overnight—and compensatory measures to improve the situation in the meantime—such as busing programs or the hiring of more black teachers and administrators for predominantly black schools—are often rejected as too expensive or too impractical. And so we have student strikes and riots, parent boycotts, and demands for wider community participation in local-government decision processes. Obviously, organizational techniques need to be developed to bring large numbers of parents and other concerned laymen into the education planning and decision-making process.

Planning exercises and problem-solving simulation games are probably among the few group problem-solving techniques that can serve these needs. Games such as those we have described for educational administrators can be exercised by concerned citizens to show how the problems of a specific community relate to the wider organization of the city or state, to show the interrelated functions of elected officials from numerous departments, to show budget limitations, and to allow for experimentation with alternative solutions to problems. Such simulations might reveal alternative courses of action not hitherto apparent and might also illustrate the structural weaknesses which constrain action by officials. In these cases, adjustments in the real world might be effectuated. In some cases, no new alternatives will result, but the citizen-players will have a greater awareness of all the forces interacting in a given situation and of the extent and limitations of the power residing in certain administrative offices.

City government is not the only form that can benefit from interactive games and simulations. In many American towns, the "town meeting" of the last century has degenerated into a backroom caucus. Games and simulation techniques could invite greater participation in community planning and decision-making. The ability of residents to manipulate simulated data concerning a proposed school, budget, or zoning change can reduce their feeling that they have little say in what goes on in town government, and improve their ability to participate constructively.

Games for community action-planning can also be useful in the analysis of social problems such as law enforcement and poverty. One of the chief problems in law enforcement is that police officers and the public frequently have different concepts of the officer's role and responsibilities. In some cases, an officer is seen by the community as having taken unjustifiably harsh actions. In other cases, he is seen as having "done nothing" and having betrayed the public trust. A reconciliation of role concepts is needed if good community-police relations are to exist. Simulation and game techniques can be developed to help policemen and citizens understand the different roles which must be played by law-enforcement officers in different situations.

A police-community relations game can be developed around an emergency involving possible conflict or cooperation among various population groups and the police. The actors would include police management, police officers, community leaders, and ordinary citizens. Citizen groups represented could vary in number and type: one set might include black militant, black moderate, white liberal, and white conservative; another might include black poor, black middle class, white poor, and white middle class. Since all players have objectives within the game context and the "rules of the game" are made more explicit in simulation than in real life, it is easier for players to understand the processes involved than when they are emotionally embroiled in the real situation. The game as a microcosm of the real world allows the players to see situations whole, rather than piecemeal as in reality.

Training games for police academies can also be developed to teach new law enforcement officers the effectiveness of alternative police methods, such as crime prevention by deterrence, physical denial, surveillance, or early warning, and crime detection through rapid responses, citizen cooperation, interrogation, and different methods of collection and analysis of evidence.

The President's Crime Commission recommended that "All training programs should provide instruction on the subjects that prepare recruits to exercise discretion properly, and to understand the community, the role of the police, and what the criminal justice system can and cannot do." * It is precisely in such areas that games and simulations are excellent educational tools. The recruits can best learn to make decisions about their roles and their powers by actually making decisions and seeing what the results are. Certainly games can never cover all possible real-life situations a policeman might have to cope with, but through careful selection, lessons of broad applicability can be readily taught.

The best use of games in police-community relations is in teaching the two groups how to communicate and cooperate with each other. Through participation in a simulation, *actual* (as well as simulated) communication can be established among various segments of the citizenry as well as between citizens and

* "The Challenge of Crime in a Free Society:" A Report by the President's Commission on Law Enforcement and Administration of Justice, February, 1967, p. 112.

police, and in this way they can learn about each others' objectives and desired means for achieving these goals. The gamed situation casts the players in less threatening positions than in real life because the actions taken in the game do not have direct real-life effects, so there is more chance for the participants to develop rapport and to understand each other. The game can be structured so that certain groups must cooperate with each other and support each other's goals if they are to achieve their own objectives. Through some role reversals in which, for example, police play citizens and citizens play police, or in which blacks and whites reverse roles, the players can gain insight and empathy with people and points of view which they might normally oppose. The appropriate channels for action can be clearly defined in the game and thus help teach citizens how they can cope with legal problems in real life. Policemen and different citizen groups can help define and make explicit the implicit role expectations which they have for themselves and other community players.

Planning exercises can greatly aid the actual community and law-enforcement decision-makers in both the making of plans and in the implementation of programs. Specifically, planning and training games can achieve direct and exciting involvement of relevant citizens in community law enforcement planning; mutual education of law enforcement officials in the preferences, priorities, and needs of the residents in affected communities; education of the residents about political, economic, and technological limits of police action; quick and inexpensive experimentation with alternative law-enforcement programs to predict reactions and to help predict effects on crime rates; identification of innovative alternatives by the creative tensions of uninhibited group problem-solving using the ideas and experiences of different relevant groups; evaluation of costs and benefits of alternative plans in terms of human response, crime, and dollars; clarification of issues and the prediction of the consequences of plans; training of law enforcement officers, management, community leaders, and average citizens in cooperative decision-making and coalition-building; and increased acceptance of police community-relations units and other programs by participation of affected population segments in the planning of the programs.

There are, of course, limits on the effectiveness of games as a planning or research device. As with any simplified model of reality, games cannot predict future events. Environmental factors or critical individuals may have been omitted from the simulation, leading to inaccuracies.

The play and the outcomes are also only as realistic as the willingness or ability of the players to play roles. Casting is an important part of simulation games; if the personalities of the players are unrealistic representations of the types of personalities filling the roles in real life, the game outcomes may not be accurate. Because game playing is highly motivating and emotionally stimulating, many emotions are aroused and some participants can become almost as agitated as in real life.

These limits, however, are actually side effects of the advantages of games, for games combine the theoretical usefulness of model-building and systems analysis with dramatic, direct participation and involvement. The models are actually improved as predictors of social outcomes because of the subjective, qualitative factors introduced by human players. Aroused emotions and hidden feelings of resistance to a program, when exposed and analyzed objectively, can be channeled toward planning and can facilitate implemention of programs.

Many of the goals and principles of community planning for law enforcement are also easily recognizable in community-action planning with the poor. The following goals can be met most effectively and realistically when the poor are involved in the planning:

Sharpening and analyzing information about local poverty problems and needs.

A broader and deeper assessing of programs and resources currently or potentially available.

Considering alternative goals, priorities, and program designs.

Deciding on goals, priorities, and courses of action which will achieve the greatest total impact on poverty in the shortest time, with the resources available.

Evaluating the actual results of program operations to capitalize on successes, correct deficiencies, and modify future plans on the basis of experience.

Simulation games can achieve these goals. Games are their own motivation, and popular involvement in them provides a method by which problems may be analyzed and courses of action proposed. The chief end of the use of simulation games in community planning is to plan *with*—not *for*—the poor. Such simulation games may have the scope of a neighborhood or a city and a time span of a month or years; problems of jobs, housing, sanitation, schools, transportation, civil rights, health and welfare, and law enforcement may be included in the game.

With contemporary society going through immense social, economic, and political change, citizens sometimes find themselves preoccupied with the phenomena of transition without a clear picture of what the transition leads to. Games provide a method of translating these phenomena into simplified terms accessible to people of ordinary education and sensibilities. We may expect an increasing gradation of games from purely frivolous to intellectually substantive to educationally serious. Eventually, decision-making exercises of ultimate seriousness may be practiced in game form. The spontaneous generation of new options will occur, and the objective evaluation of already identified options will be promoted.

As men simultaneously become more technological and more socially concerned, they will seek ways to apply the techniques and principles of science to human needs. It is imperative, therefore, that scientific concepts and techniques be understood and applied in human terms. We need a new language of action which will allow for the translation of scientific analysis into human consequences, and human attitudes into analytical modes.

Serious gaming can provide this new language. It is essentially scientific in that it is based objectively on systems analysis, quantitative model-building, and reality-testing. It also provides for human interaction, problem-solving, and dramatization. The very process of developing serious games for analyzing and/or solving social, economic, and political problems encourages an objective but empathetic viewpoint towards the various factions involved in a conflict. At the same time, it involves players far more emotionally than do graphs, charts, models, treatises, or lectures.

Already games dramatizing problems have been used in theater workshops. Viola Spolin of the Theater Games Center uses game structure as a means of freeing the individual from socially imposed behavior patterns and allowing him to react fully to the influences of his environment. Her theater games bring together actors, writers, directors, stage crew, and audience in a manner which makes everybody a player in the game. The old-fashioned barriers are removed; everyone participates in a new experience.*

Similar interactive processes have been used with some success by Synanon in the rehabilitation of drug addicts. Participants are encouraged to express themselves fully, to interact, to explore and analyze the conflicts resulting from their interactions, and to come to a fuller awareness of their strengths as well as limitations. These techniques approach formal games in their structure. Other group-therapy programs also employ these techniques.

To date, the objective of most such programs has been to give the participant a greater awareness of himself as he interacts with others. The games concept has far wider applications, however. It is not difficult to imagine a theater experience in which very large groups of people play themselves in a game and interact with others to explore and analyze a social problem. Likewise, it is not difficult to imagine large groups of people interacting and exchanging views on a political or economic problem, combining game techniques with group therapy.

* Viola Spolin, *Improvisation for the Theater* (Evanston, Illinois: Northwestern University Press, 1963).

It seems to be merely a matter of time before the principles of serious gaming analyzed in this book and used for problem-solving on the largest scale coincide with the same interactive techniques already used on the much smaller scale of individual problem-solving and analysis. If this day arrives, we will be closer to the classical ideal of life—in which thought and action, individuation and participation, are combined in the same activities. Once again citizens will be able to participate in social decision-making. Once again man will be able to enjoy a life of action. Once again man will be able to feel the zest for life which comes from enacting great changes on the models of great ideas.

APPENDICES

A

ELEMENTARY MATHEMATICS GAMES

———————

THIS SECTION describes several games which can be used to replace drill in arithmetic problems. A special deck of 168 cards is used comprised of 16 cards for each number from 0 to 9, and 8 "wild" cards which can represent any numeral.

"Equation Match" is a card game similar to "Casino" but using more complex arithmetic processes. The game is played in this manner: Each player is dealt, face down, a specified number of cards depending on the number of players (see table), and three cards are turned face up on the board. The dealer makes the first move, the dealership rotating in successive rounds. The players, in turn, combine any two or more cards on the board through addition, subtraction, multiplication, or division, or any combination thereof, to reach a number which matches one of the cards he has in his hand. The player then captures all those cards. If a player cannot take any cards, he must place a card from his hand on the board.

The following moves might occur in a game with three players. Board cards are, say, 2, 7, 9. Player A has in his hand 9, 4, 5, 1. He could add the 7 and 2 on the board and match the sum with the 9 in his hand, but he would be wise to find the better possibility of subtracting the board 7 from the board 9 to get 2, and add that to the 2 also on the board to get 4, which matches the 4 in his hand. He would then, in clearing the board, force the next player to place one of his cards on the board, and the third player also to place a card on the board. He would then have a chance to make matches himself in his own turn.

Play continues in this manner until all players have used all the cards they were dealt. Any cards left on the board remain there to be used in the following round. The next dealer then deals another hand, and the game continues until all cards in the deck have been used. If a player makes a mistake, the first opponent to identify and correct it wins all cards on the board. The winner of the game is the person who captures the greatest number of cards.

To keep the action fast-paced and challenging for the players, it is advisable to have three players, no fewer than five cards dealt to a hand, and three cards placed on the board. This allows for rapid game play and at the same time makes it possible for players to use ingenuity in setting up the board to their own advantage.

In trying to maximize the number of cards taken in each turn, players often spend some time in making their moves. The game is more exciting if time limits are imposed. At first, the teacher should allow about ten seconds per move; later, as students become more competent, he should reduce the time to four or five seconds per move. If a player does not move within the time limit, he forfeits his turn and places one of his cards on the board. This is actually a double penalty since he loses the opportunity to win cards on that turn and increases the number of possible arithmetic combinations his opponents can make.

The following game elements can be manipulated to provide for the amount of drill the teacher thinks is necessary: the number of players per game, the number of cards in the players' hands, the number of required board combinations before a player can win cards, and the number of arithmetic processes to be used. The more players, the more cards should be dealt in each hand to insure that each player will have a fair number of chances to play. The greater the number of players and cards, the greater the competition for the cards on the board. On the other hand, the more players, the less the opportunity for a player to set up the board for a future move.

The teacher can control the number of arithmetic processes which students will use in this game. The game may be played with only addition or only subtraction, but it is recommended that both be used since the number of playable combinations is greatly increased. Multiplication and division should not be used as the only processes in the game, but should be combined with addition and subtraction to make numerous combinations possible. Here are the various combinations:

Addition only
Subtraction only
Addition and subtraction
Multiplication, addition, and subtraction
All four

"Equation Rummy" is a somewhat more challenging variation of "Equation Match" which adds some elements of gin rummy. The game operates in much the same way as "Equation Match" except that each player is required to use a specified number of cards from the board in each match; two or three, depending on the players' level of skill. Each player is also required to reach as match totals all the numbers from o to 9. After each match is taken by a player, he places the match face up in front of him with the card for the total on top so that he knows what number matches he must still get to win the game. The first person who makes all matches from o to 9 wins the game. Players receive two points for each match in the 0 to 9 series. In the event that no player is able to make the complete run from 0 to 9 before the deck runs out, the winner is the person with the highest total.

Much more review of arithmetic processes occurs in this game than in "Equation Match" because players must try various arithmetic methods to arrive at a steadily shrinking list of numbers. In addition, this game increases opportunities for strategy. As players build up their matches, it becomes more and more difficult for them to receive in their hands the missing numbers they need. The teacher may choose from several alternatives which solve this problem:

1. Players may be allowed to make duplicate matches and to keep these extra matches for additional points in their score. They are given one point for each extra match.

2. Players who can make extra matches may be allowed to trade them with other players who have extra matches which they need for their run from 0 to 9.

3. Players may be allowed to make extra matches to use up some of the cards on the board, thereby putting opponents at a disadvantage. They then put the extra match in the unused deck.

Another variation of "Equation Rummy" allows players to score one extra point for each match which contains one card more than the required number. For example, if each match is required to use three cards from the board, and a player uses four, he receives one extra point. It is possible then for a player who is unable to complete his run from 0 to 9 to win over another player who does complete the run if, in a significant number of turns, the player uses more cards than required and/or is allowed to count extra matches in his total.

The object of the game "Equals" is for each player to make an equation from the cards he holds in his hand. The game is played equally well with two to four players. At the beginning of the game, at least four cards are dealt to each player. Each player looks at his cards and applies whatever arithmetic processes his teacher allows so that two cards combined equal two other cards combined, or three cards combined equal a fourth card.

The game can be played allowing addition only; addition and subtraction; subtraction only; addition, subtraction, and multiplication; multiplication and addition; multiplication and subtraction; addition and division; subtraction and division; and addition, subtraction, multiplication, and division. The game offers the greatest number of possibilities for review and is more rapidly played when all four arithmetic processes are allowed. The allowable operations with some sample equations follow:

Addition only: $2 + 3 = 4 + 1$; $6 + 4 = 7 + 3$; $4 + 4 = 8 + 0$.
Addition and subtraction: $7 + 2 = 9 - 0$; $5 + 3 = 9 - 1$; $8 + 3 = 6 + 5$.
Subtraction only: $6 - 3 = 8 - 5$; $9 - 4 = 6 - 1$; $3 - 2 = 7 - 6$.
Addition, subtraction, and multiplication: $5 \times 2 = 6 + 4$; $3 \times 3 = 9 + 0$; $5 + 7 = 6 \times 2$; $7 - 7 = 5 \times 0$.
Multiplication and addition: $7 + 1 = 8 \times 1$; $3 \times 4 = 6 + 6$; $9 \times 1 = 7 + 2$.
Multiplication and subtraction: $3 - 2 = 1 \times 1$; $9 - 3 = 3 \times 2$; $7 - 3 = 4 \times 1$.
Addition and division: $5 + 3 = 8 \div 1$; $2 + 2 = 8 \div 2$; $3 + 0 = 9 \div 3$.
Subtraction and division: $5 - 3 = 8 \div 4$; $9 - 6 = 6 \div 2$; $7 - 6 = 4 \div 4$.

When a player can make an equation on his turn, he places the cards on the board and recites his equation. If he is correct, he keeps the cards as a "book" and draws four more cards. If he is incorrect, another player may challenge him and win the book. If a player is incorrectly challenged, the challenger must give his opponent one of his books. If a player cannot make an equation on his turn, he discards one card and draws a new card. If he can make an equation using his new card, he must wait until his next turn to do so. The winner of the game is the player who has the greatest number of books when the deck has been used completely.

This game becomes substantially more difficult as more cards are required to be used in making equations. With five cards per hand, each player must think of all possible combinations before discarding. This may become a time-consuming process. If there are more than two players, however, each player will have the chance to think through his hand before he has to make a move. Some possible equations are:

$$6 \times 4 \times 1 = 8 \times 3$$
$$6 \times 4 + 1 = 5 \times 5$$
$$5 \times 3 + 1 = 8 \times 2$$
$$5 \times 5 - 1 = 8 \times 3$$

The teacher may speed up the game by imposing time limits on each move; students time each other. If a player fails to make a move in the required amount of time, he forfeits his turn. The recommended time for each move is ten seconds if there are three or four players, and fifteen seconds if there are only two players. A five-second "checking" allowance is made after each move so that other students may look for any errors made. The player who is able to figure out his hand fastest will have the advantage of making more books, but the very slow student will not be severely penalized since he can pass for any turn in which he does not have enough time to play his move and can win books by identifying errors in his opponents' equations. The slow player will quickly learn that it is better to take his time and be correct, and to check the moves of others, than it is to hurry and be wrong.

"Equals" is easily adaptable for team play, and it proceeds faster, and sometimes with greater competition, if there are two teams with two players each. In team play, each player should have no more than four cards, and preferably only two or three. Team members sit opposite each other, and play rotates between teams. The first player from Team I plays one card on the board and draws a new card. The first player from Team II does the same, setting his card separate from Team I's. The second player from Team I plays a card alongside his partner's and draws; he is followed by the second player from Team II. Thus, two sets of cards appear simultaneously on the board, one for each team. The objective of each team is to make an equation in four cards, the first two cards equaling the second two, using any arithmetic process to arrive at the numbers. Players do not tell what processes they might have in mind when playing cards. Play might go:

Team I	*Team II*
Player A plays a 5	Player B plays a 3
Player C plays a 2	Player D plays a 1
Player A plays a 6	Player B plays a 2
Player C plays a 4 and	Player D plays a 2 and
completes the equation	completes the equation
$5 \times 2 = 6 + 4$	$3 + 1 = 2 + 2$

Whenever an equation is completed, the player who completes it reads it aloud.

If the player of the fourth card cannot complete an equation, he discards a card from his hand and draws a new one. His partner then, at his turn, may complete the equation. If a player mistakenly plays a card at the fourth turn that does not complete an equation, the other team wins the book by naming any number needed to complete an equation.

The possibilities for completing an equation are numerous. Player C of Team I could have played other numbers than the 4 to complete an equation. He could have played a 1 ($5 + 2 = 6 + 1$), a 2 ($5 - 2 = 6 \div 2$), or a 3 ($5 - 2 = 6 - 3$). If he had played a 5, however, the book would have been sacrificed to Team II since 5 does not create any equation with 5 and 2 on one side and 6 on the other.

Numerous variables can be manipulated in team play. Equations are far more difficult to complete if there are only two cards to a hand than if three or four cards are used. In addition, the game becomes more difficult and requires much more thought if equations using five rather than four cards are required. Since no player knows what arithmetic processes his partner has in mind when playing a card, each player will be forced to think through all the possibilities for the equation on each turn he makes. The player who supplies the final card of an equation reads it aloud for the other team to check. A significant advantage which team-play of five-card equations enjoys over four-card equations is that the lead for each team is likely to alternate. No single player, then, will be forced to complete each equation each time, which could happen with four-card equations if a player had the right cards for each final move.

"Multiplication" and "Division," two games designed to offer practice in the structuring of problems in multiplication and long division, are more drill-oriented than the previous games. They are also more challenging. Both games require a high degree of conceptualization on the part of students, but they can be played on a number of levels. The multiplication game is the less complex of the two since it requires only multiplication and addition as processes; the division game requires division, multiplication, subtraction, and addition.

Both games may be played by two to four players. To start either game, players draw cards randomly from the deck to set up a multiplication or division problem. On its simplest level the multiplication game employs a one-card multiplicand and a one-card multiplier. On its most complex level, the game employs a three-card multiplicand and a two- or three-card multiplier. The long-division game, on its simplest level, has only one card in the divisor and three cards in the dividend. On its most complex level, two cards are used in the divisor and five cards in the dividend.

After a problem has been set up such as 32×7, or $563 \div 7$, players are dealt a fixed number of cards (see tables on page

144). Each player in turn attempts to place one card anywhere it belongs in the solution to the problem. After he has done so, he places a marker on the card to identify it as his, and he draws a new card. If a player cannot place any of his cards in the problem, he places one of his cards face up in front of him saying, "This number does not appear anywhere in the problem," and draws a new card. If he has a number that has already been played in the problem, he places his card beneath the played card but does not put a marker on it; he then draws a new card.

If a student makes a mistake in placement, an opponent on his turn may challenge the placement by turning the card over and putting the original marker button beneath it; the challenger then puts his own marker button on top of the card to indicate that he has identified an error. If the challenger has the correct card in his hand, he places it on top of the overturned one and places an additional marker button on top. If he does not have the correct card, he makes a move of his own.

When the problem is solved, players count up their scores, counting one point for each marker button on a correct card and subtracting one point for each marker under an incorrect card. They then check the cards they have said do not appear anywhere in the problem. They score one point for each card they were correct in saying could not be played; they subtract one point for each card that could have been played.

The teacher may wish to use large graphlike charts to help students visualize the problems at the early stages of the game. This will help them avoid placing a correct number in the wrong column. Examples of guiding charts are on page 144.

As students gain proficiency in multiplication and long division, they will be able to avoid using the chart, thereby making the game more difficult and more competitive. The number of cards used in setting up a problem affects the number of players recommended and the number of cards used in each hand. The simpler problems should have no more than two players; these players should have four or five cards in their hands to make

```
        9 7 2
          4 6
      5 8 3 2
    3 8 8 8
    4 4 7 1 2
```

```
              8 7 2
      8 3 ) 7 2 3 9 1
          6 6 4
            5 9 9
            5 8 1
              1 8 1
              1 6 6
                1 5
```

problem-solving rapid and less dependent on the chance drawing of the "right" cards. More complex problems are easier and
faster to solve with four players, each having three cards in a
hand. The limitation of the number of cards requires each student to think through, to the best of his ability, the entire process involved in the problem. Longer problems offer a greater
chance of using a given number in the answer, because many
more digits occur. Here are the recommended distributions:

MULTIPLICATION

Number of Players	Cards per hand	Cards in Multiplicand	Cards in Multiplier
2	4 or 5	1	1
2	4 or 5	2	1
2	4 or 5	2	2
3 or 4	3	3	1, 2, or 3

DIVISION

Number of Players	Cards per hand	Cards in Dividend	Cards in Divisor
2	4 or 5	3	1
3	4 or 5	4	1
3 or 4	3	4 or 5	2

The "Multiplication" and "Division" games offer several variations. For example, four players can play as teams of two players each. The members of a team are allowed to talk over moves with each other and to decide on a single move. One good student matched with a poorer one gives him special, individual help when the two discuss and play their moves.

Also, the teacher can select digits to appear in the multiplier or the divisor in order to concentrate on numbers the students have difficulty with. If, for example, the students have trouble with the tables of 8 and 9, the teacher may require that a multiplication problem have a multipler of 89; a division problem might have a divisor of 89 or 98.

Another variation is to reduce the number of cards in each hand to two and change the rules so that both cards must be played at once. Whenever a player cannot play both cards anywhere in the problem, he turns in one card and draws a new one; this process continues until one player can place both cards. Students then recite the problem orally to check whether the player is correct. Each time a player correctly places two cards, he scores one point. A class winner can be declared at the end of the class.

Once students have acquired skills in basic arithmetic and have mastered the basic concepts of the games, the range of variation is enormous. Students, having mastered the easier forms of the games, tend to concentrate on the more difficult versions since they are more challenging, more interesting, and more exciting. Because winning the game is somewhat important to the players, each will try to make the game as difficult as possible for his opponents.

One version which acquires elements of complex strategy is to have each player specify a particular arithmetic operation (addition, subtraction, etc.) which the next player must use.

Students should be encouraged to experiment with the variables for each game themselves. In some cases, their changing one variable will make the game easier; in some cases, more difficult; in some cases, unplayable. Students will gain much from this experimentation—even from the experiments which fail— since they will be engaged in divining the mathematical relationships of the variables involved, and this itself is a sophisticated activity different from the practice of arithmetic.

B

A GAME FOR PLANNING
AN EDUCATIONAL SYSTEM

The education-system-planning game described below was played by participants in the Conference on Educational Innovations held at Lake Arrowhead, California, in 1965. Its objectives were to illuminate some major issues of education planning, to excite an increased awareness of alternative plans and their costs and benefits, and to stimulate a problem-focused exchange of ideas among players having diverse approaches to education.

At least twenty players are needed, divided into two educator teams, two student teams, and one team of "reality daemons." The educator teams include roles representing several levels of the educational establishment. The student teams are of two types, advantaged and disadvantaged. The reality daemons team includes roles personifying various social problems that affect education planning. The complete cast of player roles is given on page 148.

The objectives of these five teams are to exercise either their

Player Roles in Education-System-Planning Game

	Educator Team A	Student Team A	Reality Daemons 1 to 8
A1	U.S. Commissioner of Education	A1 Advantaged elementary students	1. Educational establishment
A2 *	State Commissioner of Education	A2 Advantaged high-school students	2. Public-opinion acceptance
A3 *	Board of Education chairman	A3 Private-university students	3. Economic productivity
A4	City-school superintendent	A4 Adults with college degrees	4. Intellectual creativity
A5	Elementary-school principal		5. Psychological satisfaction
A6	Elementary-school teacher		6. Regional upgrading
A7	High-school principal		7. Political participation
A8	High-school teacher		8. Social-conflict reduction
A9 *	University president		
A10 *	University professor		

	Educator Team B	Student Team B
B1	U.S. Commissioner of Education	B1 Disadvantaged elementary students
B2 *	State Commissioner of Education	B2 Disadvantaged high-school students
B3 *	Board of Education chairman	B3 State-university students
B4	City-school superintendent	B4 Adults without college degrees
B5	Elementary-school principal	
B6	Elementary-school teacher	
B7	High-school principal	
B8	High-school teacher	
B9 *	University president	
B10 *	University professor	

* Roles A2, B2, A3, B3, A9, B9, and A10, B10 may be eliminated if an insufficient number of players is available.

education-system-planning skills (educator teams) or their edu-
cation-evaluation skills (student teams and reality daemons)
and to identify obstacles and problems. There are three simultaneous competitions—between educator teams, student teams,
and reality daemons. The team objectives are summarized below.
(Note that all the game objectives are also important objectives
of any education-system-planning and programing effort in real
life.)

Educator Teams A and B compete with each other to achieve
within a given budget the greater "net educational product"
within constraints imposed by the student populations and the
reality daemons. Net educational product (NEP) is defined as
the number of graduates of elementary, secondary, and university schools weighted by their "quality." Quality is determined
by evaluations made by each of the reality daemons, measured
on a one-to-ten-point scale.

Student Teams A and B compete with each other to achieve
improvement in the greater net educational product, as a result
of using Educator Team A's or B's plan. Improvement is measured by the ratio of net educational product following implementation of an educator team's plan, to initial net educational
product.

Reality Daemons 1 to 8 compete with each other to discover
the largest number of implausibilities in the "gross-educational-
product" estimates which the student teams made of the educator teams' plans. Each of the reality daemons seeks to be the
principal source of reductions from gross to net educational
product. In this game, reality is actively malevolent, and the
reality daemon who convincingly disallows the most gross educational product wins. To discourage arbitrary and unrealistic
disallowances, and to obtain a record of disallowances useful in
postgame analysis, every disallowance must be justified by a
one-page written argument.

The sequence of simulated activities is as follows: In the first
hour, the educator teams formulate their plans under the needling of the reality daemons. This needling takes the form of in-

puts of troublesome social and educational problems that must
be dealt with by the educator-planners, who have initiated their
planning efforts with only a few "reality problems" supplied by a
scenario. The formal product of this planning session is a filled-
in planning form (below) with backup explanations. This consti-
tutes the team's national education plan.

EDUCATION-SYSTEM-PLANNING FORM

| | | EDUCATION-SYSTEM COMPONENTS | | | |
		Curricula	Teachers	Media and Equipment	Facilities
Advantaged	Elementary (K–8)				
	High school (9–12)				
	Private university				
	Adults with college degrees				
Disadvantaged	Elementary (K–8)				
	High school (9–12)				
	State university				
	Adults without college degrees				

At the same time, the reality daemons give the student teams
social and educational problems, but from a more personal point
of view. The student teams will use these problem inputs to prepare a set of criteria for evaluating the national education plans
they receive in the second hour.

At the end of the first hour, both educator teams submit their national education plans to both student teams. During the second hour, each student team—one advantaged, the other disadvantaged—evaluates the two competing national education plans on the basis of their probable educational contribution to their respective populations. Each student team chooses one of the plans for implementation (they may both choose the same plan). Each team then estimates the impact of the plan on its population, on the basis of improved quantity and quality of educated individuals. This improvement is stated as a gross educational product (GEP).

In the third hour, these GEP estimates are submitted to the reality daemons, who eliminate or adjust implausible claims for improved GEP, thus producing a net educational product (NEP). The educator team with the largest NEP wins.

On page 152 is a table showing the three stages of the game.

Before the start of play, each player is given a microscenario of the role he will play. Here are some sample microscenarios for members of the *Advantaged student team:*

You represent the *elementary-school* population of the *urban and suburban middle class.* Family income ranges from $6000 to $20,000 per year, with a median income of about $10,000. Most students have at least one parent who completed college. Almost all heads of families are white, American-born, and employed.

You represent the *high-school* population of the *urban and suburban middle class.* Family income ranges from $6000 to $20,000 per year, with a median income of about $10,000. Most students have at least one parent who completed college. Almost all heads of families are white, American-born, and employed.

You represent the *private-university* school population of the *urban and suburban middle class.* Family income ranges from $6000 to $20,000 per year, with a median income of about $10,000. Most students have at least one parent who completed college. Almost all heads of families are white, American-born, and employed.

SEQUENCE OF ACTIVITIES FOR
EDUCATION-SYSTEM-PLANNING GAME

	2 Educator Teams	2 Student Teams	Reality Daemons Team
First Hour	Each team formulates national education plan and budget allocation, using planning form, scenario input, role assignments, and education problems from reality daemons.	Each team prepares criteria for evaluating desirability and estimating consequences of alternative plans, based on own population characteristics.	Give social and education problems to educator and student teams.
Second Hour	Clarify and elaborate on their plans, at the request of the student teams for more information.	Advantaged and disadvantaged student teams each choose most desirable national education plan, estimate improved gross educational product resulting from plan.	Continue to introduce social pressures and life problems to the two student teams.
Third Hour	Defend and explain their plans to reality-daemons jury.	Defend and explain their GEP estimates to reality-daemons jury.	Jury of reality daemons determines net educational products of competing plans, and on this basis the "winning" plan.

NOTE: Heavily outlined activities are the principal arenas of decision-making in the game.

And for the *Disadvantaged student team:*

You represent the *elementary-school* population of the **urban disadvantaged**. Family incomes are mostly under $6000 per year. More than half are Negro or Puerto Rican. Many heads of families are women with part-time jobs, and most heads of families have not attended college. Many high-school students must work at least part time to support their families.

You represent the *elementary-school* population of the *rural dis-*

advantaged. Most family incomes are less than $4000 per year. Most parents have not finished high school. Many are Negroes, Mexicans, or Indians. Most parents are small farmers, agricultural workers, or miners, and families are large.

Here are the starting instructions given to educator teams:

Read the *Player Roles,* statement of *Team Objectives,* and the 3 x 3 table, *Player Sequence of Activities.* Examine the blank *Education-System-Planning Form.* Your team has one hour to complete this Planning Form and a half page of written backup explanations for each of the 32 budget and descriptive categories (e.g., "Disadvantaged High School—Facilities"). The completed Planning Form and backup explanations constitute your team's National Education Plan, and is immediately submitted at the end of the hour to the two-hour Student Population Teams. The Student Teams also receive a competing National Education Plan from your competitor, the other Educator Team. If at least one of the Student Teams chooses your team's plan, and if it "performs" better than the other plan in terms of Net Educational Product (NEP) in the evaluation by the Reality Daemons, your team "wins."

To give you a basis for beginning the planning effort at once, assume the current (1965) state of U.S. education and its attendant problems. There are pockets of poverty, racial tensions, rising crime rates, and backward regions. A budget will be given to you, for each of the four education-system components, based on the Delphi Procedure. This is a guide which you may modify in individual components, but the total must remain the same. Remember, however, that you have only one hour, and must describe each of the 32 budget categories on the table *substantively* in a half-page handwritten explanation.

To speed your planning effort, the player in the role of U.S. Commissioner of Education will act as team chairman and make all final decisions on choices. It is suggested that one way to assure completion of your plan in time is to divide the labor by assigning players in specific roles to specific planning functions (e.g., high-school principal to planning high-school curricula, teachers, equipment, and facilities).

Here are the starting instructions for student teams:

Read your pink or blue student *Population Scenarios,* the *Player Roles, Team Objectives, Player Sequence of Activities,* table, and *Education-System-Planning Form.* Your team will receive two competing *National Education Plans* from the two Educator Teams at the end of the first hour, consisting of the completed *Planning Form* and about 15 pages of written description of each of the 32 budget categories (such as "Disadvantaged—Elementary—Curricula"). Your problem is to evaluate the two competing plans in terms of what they do for your populations in terms of Gross Educational Product (numbers of graduates of various levels of x quantity). Selection of the most desirable plan must be by majority vote in your team. After selecting what appears to be the most desirable plan for your particular mix of student populations, write a two-page description of the improvement in Gross Educational Product (GEP) you expect as a result of the implementation of the selected National Education Plan. Your estimate of the improved GEP must be completed at the end of the second hour and submitted to the Reality Daemons for their evaluation. They will reduce your expectations by eliminating what they consider implausible or impractical, and determine a Net Educational Product (NEP). If your team's improvement in NEP is greater than that of the other student team, your team "wins."

During the first hour while waiting for the National Education Plans to be submitted, you will be receiving various life problems from the Reality Daemons. Try to imagine and note down how your particular segment of the student population reacts to these problems in the current educational context so that when you see the new plan you may more readily estimate its effects on your population.

In the game played at the Lake Arrowhead conference, the two educator teams developed quite different National Education Plans. Educator Team A, which included a real-world United States Assistant Commissioner of Education, allocated most of its ten-billion-dollar budget to training additional teachers for elementary and secondary schools in disadvantaged areas. Team B, which included a real-world Director of Manpower Studies of the National Science Foundation, distributed its ten billion dollars much more broadly, across both advantaged and disadvantaged populations, and for facilities and equipment as well as for more teachers and more teacher training.

As might be expected, the advantaged student team chose the latter plan, while the disadvantaged student team chose Plan A. Neither student team was really happy with either plan, finding many gaps in planning that were corrected only partly by direct consultation with the educator team planners. If more time had been available (days rather than hours) the student teams might have rejected both plans and insisted that the educator teams repeat their efforts and formulate more satisfactory plans.

The reality daemons jury reduced considerably the students' GEP estimates to NEP's. They then estimated the increased number of high-school graduates each plan would produce; Plan A was a clear winner here, with its emphasis on teachers for disadvantaged schools with previously high dropout rates. The quantity of graduates then had to be weighted by their estimated increased educational quality. Plan B, with its more balanced distribution of improvements, clearly produced a better improvement in average quality. However, its qualitative superiority was just barely insufficient to overcome the quantitative superiority of Plan A, according to the relative weightings given quantity and quality of educational output by the reality daemons. Thus Plan A won.

While the plans produced by this particular abbreviated game were much too uncoordinated to be directly applicable to a real school system, the issues discussed and the apparent consequences of budgetary decision offered a dramatic and intensive problem-solving experience to the participants. The game motivated a much more problem-focused and solution-stimulating activity than had an earlier discussion of the same planning problems. The participants, through role-playing and making joint decisions, were stimulated to produce more issues, arguments, and plans than they had done otherwise.

This game can be played by any group of education planners that wish to make the effort. No special equipment, facilities, or skills are required. All that is needed are twenty or more serious players who have at least three hours to spend.

C

"SEPEX": A SCHOOL ELECTRONICS
PLANNING EXERCISE

THE "SEPEX" game was designed to aid educational decision-makers in the application of electronic systems to educational services, both instructional and administrative. The game employs simulation and role-playing techniques to create understanding of the feasibility, the potential educational benefits, and the costs of alternative electronics systems for interconnecting school districts in large geographic areas of low population density. As a simulated planning process, it induces participants to consider how electronic systems could improve the efficiency of current educational sources, how these systems could lead to desirable and feasible future services, and how their cooperative use might aid in increasing and equalizing educational opportunities among various districts.

"SEPEX" is played by fifty to sixty players over a three-hour period in an area consisting of five adjacent rooms. It requires about an hour's preparation, including briefing, and should be followed by at least an hour's discussion of results. A convenient schedule would be as follows:

10–11 a.m.: Explanatory briefing
11–12:30 p.m.: Game play (first half)
12:30–1:30 p.m.: Lunch by team groups to permit strategy discussions
1:30–3:00 p.m.: Game play (second half)
3–4 p.m.: Scoring and game analysis
4–5 p.m.: Discussion of results

Setting of the Game: The game simulates the partly cooperative, partly competitive educational-planning activities of three adjacent county-school districts in the Onrush Peninsula of the north-central state of Nagitchmi. The counties are:

Bay County (richest, most populous, industrialized)
Farmfield County (well off, dispersed farming population)
Woodland County (economically depressed lumber and fishing area, sparsely populated).

The Players: The fifty to sixty players are distributed among three roughly equal teams, each team representing educators and other interested parties in each of the three county-school districts. Each team is initially organized into a planning committee comprised of the superintendent and the school board, and four subcommittees to establish requirements for curricula, number of teachers, facilities, and administrative needs; these use the remaining team members.

There are twelve roles for players, distributed among the three teams, as shown on page 159.

Sequence of Activities:
1. Each of the three teams is given its educational problems in the form of planning requirements, subcommittee reports on present and future teacher needs, curriculum needs, facilities

Distribution of Players

Role	Power Points	COUNTY DISTRICT TEAMS		
		Bay County	Farmfield County	Woodland County
Public-school superintendent	8	1	1	1
Parochial-school superintendent	8	1	1	1
School-board member	10	3	3	3
Teachers (leaders)	4	3 (El., jr. hi, sr. hi)	3	2 (El., sec.)
Students	1	2 (1 mid. cl., 1 poor)	1	1
Parents (PTA leaders)	4	2 "	2	1
Farmers	2	1	2	1
Business and professional men	3	2	1	1
Union	4	1	0	1
Press/Radio/TV	4	1	1	1
Legislators	10	2	1	1
State superintendent of instruction	10	1	0	0
TOTAL PLAYERS PER TEAM		20	16	14

and equipment needs. The planning-requirements subcommittees are staffed by the players simulating the consumers of education: teachers, students, parents, farmers, businessmen, professionals, and unions.

2. Each team is given a technological menu of electronic communications, display, storage, and retrieval, and processing hardware, along with information about capabilities and costs.

3. School boards and superintendents of each country match their education requirements with the hardware menu, while considering costs and trade-offs, and make a cost-benefit comparison of alternatives.

4. Using previous steps, school boards and superintendents form a priority list of allegedly most efficient hardware combinations (systems) meeting the education-system requirements.

5. Planning groups obtain approval of own teams by vote of all team members, weighted according to power points.

6. The three teams negotiate with each other to attempt coordinate implementation.

7. One or more coordinated cross-county plans (over two or three counties) are formulated.

8. All fifty players vote on alternative plans; if all plans are rejected, they return to step 5; if that fails, to step 4.

9. If one or more plans pass plenary vote, they are scored for highest educational cost-effectiveness and maximum utilization of electronic technology.

Here is one scenario for a game of "Sepex."

The state of Nagitchmi, known as the "Moosehead state," is in the north-central group of the United States. It consists of the Onrush Peninsula, bounded by Lake Onrush on the north, east, and west; and Lower Nagitchmi, immediately south of the Peninsula. The state is twenty-fifth in size with an area of 54,221 square miles, of which 1001 square miles are inland lakes. The name of the state derives from one of the old Indian tribes of the region, the Nagichimacs, transformed by the pest-ridden early settlers into Nagitchmi.

The Onrush Peninsula consists of rugged wooded hills (Woodland County) and fertile coastal plain (Farmfield and Bay Counties).

The population of the Peninsula is approximately 1,000,000 (about one-fifth of the state's). Bay County is located on the southeast part of the peninsula. Bay County's manufacturing cities of Bayview, Wanagas, and Centerville give it the leading county population of about 500,000. Farmfield County, as the name implies, thrives on dairy and mixed farming in the central area. Farmfield County has a population of approximately 300,-000. Woodland County in the north of the Peninsula is most sparsely populated with about 200,000 inhabitants. The population has been growing at about two per cent annually in Bay County, one per cent in Farmfield County, and has remained relatively constant in Woodland County. The average population per household in the Peninsula has declined from 3.6 in 1950 to 3.4 in 1960.

The government of the state of Nagitchmi consists of a governor and a unicameral legislature. The governor and state superintendent of public instruction are elected for two-year terms. The legislature consists of a senate of a hundred members, elected to terms of two years from each district.

EDUCATION. Nagitchmi pioneered in what has become the typical American pattern of free, tax-supported and state-controlled schools. Local control through elected school-district boards was a feature borrowed from New England, but the state superintendent of public instruction as the elective head of the system was a novel adaptation from European ministries of education. County school commissioners were added later, as was a state board of education responsible for state teachers' colleges and issuance of teachers' licenses and certificates acting chiefly through the state superintendent. There are approximately ten school districts—a great decline from nearly seventy earlier in the century, resulting from consolidations prompted by efficiency considerations.

(There are three school districts in the Onrush Peninsula, corresponding to the three counties.)

The executives of the county school districts are county school superintendents, paid by the state. The superintendents are appointed by the elected county school boards.

Enrollment in the public schools in the Onrush Peninsula is approximately 300,000 or 87 per cent of the school census (ages five through nineteen). Of these, 200,000 were in elementary grades (K-6) and 100,000 in secondary (7-12).

The total public school expenditure for the school year for the state was $400 million, of which $220 million went to 40,000 teachers. In the Onrush Peninsula, it is estimated that annual school expenditures total $80 million, of which some $50 million went to 10,000 teachers. Average expense per attending pupil in the state was $300, and the average teacher's salary in the state was $5,000. The average expense per attending pupil in Onrush Peninsula was only $200, however, and the average teacher's salary only $5000. This lag behind the more industrialized southern part of the state is a source of chagrin and aspiration for the Peninsula educators.

Parochial schools accounted for about one-third of the above enrollments.

ECONOMY. The prosperous manufacturing centers of Bay County and rich farm lands of Farmfield County present a contrast to the relatively depressed lumbering, fishing, and tourist industries of Woodland County. Woodland County educators have sought help from their adjacent county colleagues in improving the employability of their students through a variety of enrichment, counseling, and vocational programs, but little progress has been made. The more conservative policies of the Farmfield County school board have inhibited the necessary expenditures, while Bay County has its own serious problems of a growing school population and technological unemployment.

POLITICS AND PUBLIC OPINION. The state has been typically middle-of-the-road in its national political preferences, with a somewhat more conservative leaning in the farm areas in local politics. Bay County industrial towns are the source of moderately progressive attitudes. The press reflects rather than forms the predominant public opinion in favor of good education for all, coupled with reduced taxes. Innovations in education are welcomed when they do not cost money or trespass on cherished religious beliefs or cultural mores. Nevertheless, many thoughtful citizens, liberals and conservatives alike, in all three counties are becoming increasingly concerned over the lag in the quality and quantity of education offered Peninsula students in terms of the rest of the state and the needs of the future. Some of the better informed educators have argued for the introduction of modern electronics communications, processing, and display equipment to increase the efficiency of instruction and administration in the schools, and this has become an outstanding issue in the state.

Committees have been formed in all three Peninsula counties to consider current and future needs for teachers, curricula, facilities, and administration so that the potential uses of electronic communicatons, monitoring, displaying, processing, storing, and retrieval equipment can be evaluated.

D

"COLONY":
A SECONDARY-SCHOOL
GAME

THE GAME of "Colony" was designed to illustrate relationships
between the American colonies and Great Britain. The game
should help students understand the difficulties the British faced
in obtaining revenue needed to pay for the defense and administration of the colonies, and also the colonists' irritations over British taxes on trade.

The game is set in the period between the end of the French
and Indian War and the American Revolution, but there is no attempt to reproduce specific historical events or to follow chronol-
ogy precisely. The British need for revenue and the American
annoyance at taxes persist throughout the period. The game centers around the basic mercantilist concept of a colony; that is,
colonies were supposed to exist solely for the economic benefit of
the mother country, supplying her with raw materials and providing a market for her manufactured goods. In return, the
mother country provided government and protection.

In spite of her mercantilist ideas, Great Britain had followed a policy of "salutary neglect," leaving the colonies more or less alone for nearly a hundred years prior to the period in which the game is set. After the French and Indian War, Britain was faced with increased costs of administering North America and with a huge war debt. She could no longer afford to ignore the colonists' evasion of payments. Hence, old trade regulations began to be enforced and new ones were passed. The colonists, long accustomed to ignoring trade restrictions, regarded their enforcement as unjust.

The colonists had created additional problems for the mother country by expanding their settlements along the frontier and developing some of their own industries. Frontier expansion led to conflicts with the Indians that interrupted the profitable fur trade and forced Britain to spend more money on defending the frontier. Colonial industries began to supply the colonial market with local goods, thus competing with English industries. The expansion of the frontier and the development of colonial industry made the colonies both more expensive and less profitable for the mother country.

While the game of "Colony" focuses on the economic conflict between Britain and the colonies, it is not intended to show that the American Revolution had purely economic causes. Few Americans were suffering economic hardship from the enforcement of trade regulations. The game should make it clear that merchants who paid taxes still made good profits. The economic issues did, however, raise the fundamental question of who ruled America. While the political aspect of this issue is not included in the game of "Colony," it can be discussed after the game has been played.

PLAYERS. The class should be divided as follows:

British government team: Four or five players. These may perform collectively or may divide up their various tasks of deciding on a tax schedule each cycle, collecting taxes, and attempting to catch smugglers by sending customs officials, warships, and judges to America.

Merchants: Three equal teams of colonial merchants divided among northern, middle, and southern sections of the colonies.

Five bankers assigned as follows: One at the British Trade Bank; one at the European Trade Bank; one at the American Import Bank; one at the American Export Bank; and one at the Investment board. Bankers buy and sell goods to merchants, handle sales of investments, and pay profits on these investments. The British Trade Bank pays bounties for the British government.

Layout and Sequence of Action

Seven game boards are set up in different parts of the room. Each merchant starts by buying goods in America. Each team then proceeds to sell goods in Britain or in Europe and to purchase goods to bring back to the colonies. Whenever a merchant team returns to the American coastline most players stop what they are doing and observe the action, since smuggling is fun.

When the boards are set up, each merchant receives £600 and an Action Schedule, which contains lists of available cargoes, a place to mark the cargo purchased, and a place to mark the smugglers' cove or customs port the merchant is returning to. A brief description of the game, including the objectives of the British government and of the merchants, should be made before play begins. Rather than make this a lengthy introduction, a demonstration move using four players (one British government player and one merchant from each of the colonial sections) is helpful in showing students what to do first. Then they can refer to their individual set of rules as necessary in the course of play.

COURSE OF PLAY. In the game of "Colony," two teams play according to different rules. The British government team has the objective of obtaining enough money from the colonies to pay the cost of governing them. To achieve their objective, they may tax various British commodities traded with the colonies and allocate administrative resources, in the form of warships, customs officials, and judges to enforce their rules. The use of these resources costs the British money. Prices of each are marked on the game board.

The British are at a disadvantage. They must expend more and more money in the attempt to obtain revenue. In historical fact, it cost the British about four times as much to maintain the customs administration in North America as they obtained from duties on trade.

The colonial merchants have the objective of making as much money as possible. They export raw materials from the colonies and sell them to Great Britain. Then they buy manufactured goods in Great Britain and sell them in North America. They may, however, decide to buy and sell goods in foreign ports where prices may be better, but they run the risk of getting caught while involved in this illegal trade. When the colonial merchants return to North America with their cargoes, they must decide whether to pay the duties set by the British or smuggle their goods in through inlets and coves. Once they decide through which cove they intend to smuggle their goods, players mark the number of the cove, as shown on the American coastline board, on their Action Schedule. Meanwhile, the British indicate on their Strategy Sheet the points at which they are going to station either customs officials or warships. Should a smuggler get caught by a warship, he is fined £300 per ship. Should he get caught by a customs official, however, several outcomes are possible. He can choose a Mob card and, if he is lucky, a sympathetic mob will scare the official away. Or he can choose a Bribe card to see whether the official will accept a bribe. If the official is honest, the smuggler will have to go to court. Here three outcomes are possible. He may be acquitted by a jury, fined lightly by a jury, or tried in the Admiralty Court by a judge where a stiff fine is certain. These outcomes are represented on Court cards. The British team, if they find that juries are not convicting smugglers, may send more judges, in the form of Judge cards, which they add to the deck of court cards. In the course of the game, merchants should find that the odds of getting caught smuggling are in their favor. The British will have difficulty in obtaining the sums they need because in the process of making smuggling less economical, they expend large sums of money.

At this point, the merchants would have either:
1. traded legally and paid the duties, or
2. smuggled and gotten away with it, or
3. smuggled and gotten caught by a warship, then been fined heavily, or
4. smuggled and gotten caught by a customs official who
   a. was rendered ineffective by a mob, or
   b. was bribed, or
   c. succeeded in getting the smuggler to court where he was
      i. acquitted by a jury, or
      ii. fined by a jury, or
      iii. fined heavily by a judge.

The game is not merely about smuggling, however. Before buying goods in North America to sell abroad, players may invest in land, industries, or more ships. If they invest in industries, they get a fixed income every cycle. Ships, represented on the Action Schedule, may be purchased to increase a merchant's volume of trade. Of these investment possibilities, only domestic industries violate the rules of the imperial relationship, because domestic production reduces the market for the British manufactured goods. In fact, in the game, the price of iron in Great Britain is higher than the price a merchant could get for selling it in North America where a great deal of iron is manufactured. Since, from the merchants' point of view, it would be foolish to import iron, the British iron industry suffers. There is little the British can do. In reality, they passed a law in 1750 forbidding the building of new iron foundries, but the law was ignored.

After investments have been made, players resume trade by buying goods in North America for shipment abroad. For purposes of simplicity, the price of these goods remains fixed, except in the case of fur. Players take a Chance card to find the price

Action Schedule

Each cycle in the game of "Colony" consists of the following nine steps:

	Northern merchants	Middle merchants	Southern merchants	British government
1.	Buy goods overseas	Buy American goods	Invest	Decide how many customs officials to send and to which coves, and pay for them.
2.	Return home	(Observe)	(Observe)	Collect taxes. Catch smugglers.
3.	Sell goods in America	Sell goods overseas	Buy American goods	Decide how many customs officials to send and to which coves, and pay for them. Pay bounties.
4.	Invest	Buy goods overseas	Sell goods overseas	Pay bounties.
5.	(Observe)	Return home	(Observe)	Collect taxes. Catch smugglers.
6.	Buy American goods	Sell goods in America	Buy goods overseas	Decide how many customs officials to send and to which coves, and pay for them.
7.	(Observe)	(Observe)	Return home	Collect taxes. Catch smugglers.
8.	Sell goods overseas	Invest	Sell goods in America	Pay bounties. Plan taxes for next cycle.
9.	All merchants collect money from investments. Repeat cycle.			Write new taxes on chart.

of fur at any moment in time. Perhaps an Indian war is raging, making it impossible to get fur. The British must send troops to end the trouble. During an Indian war, the British can restrict investment.

As the game progresses, the portion of the board representing North America develops as industries are built, land is settled and worked, trade expands, and the general wealth of the colonies increases. The British try to reap some benefit from the increase, but the British are doomed to fail. The old system of taxing trade for the purpose of regulating its directions was inadequate when used for the new purpose of providing a source of significant revenue. Fundamental alterations in the imperial relationship were required, but British statesmen and political theory were inadequate to meet the challenge.

The fact that they are doomed to fail is not immediately apparent to students on the British team. They may become more frustrated in the course of the game, but this should not impair the enjoyment of trying to exercise authority in the context of the game. They may experience some degree of success, finding that merchants will pay taxes if the rates are competitive with the cost of smuggling.

The winner of the game for the colonists is the player who has the most money at the end. The British team simply gauge their success on whether or not they end the game with more money than they began with. It is unlikely that students will care very much who wins the game once they become involved in playing it. More important is the discussion following the game, which reveals and reinforces what students have learned through it.

POSTPLAY DISCUSSION. The simplest way to begin this discussion is to consider the courses of action the various players took. What did the merchants trade in? What kind of profits did they make? Did they smuggle or pay the customs or both, and why did they decide to do one or the other? What other investments did they make, and were they profitable? What did the British

do? How did they view the colonies? Did they use all their resources? Did that help to bring in a lot of revenue? Or did it
cost more than it brought in? Did they find that higher or lower
taxes were more successful in getting the colonists to pay? Or
didn't it matter? How did the colonists feel about British regulations? Did they really represent an obstacle to trade?

A comparison between what occurred in the game and what
occurred in reality should be fruitful. If the British taxes did not
really represent a hardship, why did the colonists complain
about them so bitterly? If the British really did need money,
why weren't the colonists willing to pay their share? Here is an
opportunity to get behind the conventional economic explanations of the American Revolution as well as the notion that the
British were merely oppressive rulers, out to wring from the
colonists all they could get. The fact that each side in the conflict
viewed the issues differently is important not only in this specific
case but in attempting to understand any conflict.

If the economic arguments are insufficient to explain the Revolution, what about the political? Were the colonists right in
their claim that the British were taxing them illegally? This question and others like it are not raised in the game itself but would
be valuable to introduce in the postgame discussion. No single
answer is sufficient. Students should learn to deal with a variety
of explanations and accept ambiguous solutions.

INTRODUCTION

"Colony" is a game about trading conditions in the American
colonies between 1763 and 1775. There are three roles to play in
the game: colonial merchant, member of the British government,
and banker. Each part has its own rules.

RULES FOR MERCHANTS

Goals. Your home is in the Northern, Middle, or Southern colonies. By trading and investing, you try to make as much money
as possible. The merchant with the most money wins the game.

Action. You start with one ship, £600, and an Action Schedule.
On the Action Schedule are your Merchant Fleet, Cargo Lists,

Destination Chart, and Land Ownership Chart. In each round there are six steps. The Action Schedule tells you what to do at each step.

Step 1: Buying American Goods. The top part of your Cargo List shows the goods you can buy in your home region and the price per load of each. A ship holds three loads of goods, either the same or different goods. On your Cargo List, mark which goods you are buying. Pay the American Export Bank. Try to buy goods that will give you a good profit.

(*Fur:*) If you want to buy fur, draw a Fur Price chance card. The card gives the price of each load of fur. Fur prices can be high or low. An Indian war at Step 6 stops fur trading for that round.

(*Bounty goods:*) Britain especially wants certain goods. The government will pay a bounty of £20 for each load of these goods. Bounty goods are starred on the Cargo Lists.

Step 2: Selling Goods Overseas. Decide where to sell your American goods. The British government has laws against selling most goods in Europe. But you may decide to sell goods illegally if you think you can make more money and will not get caught. If you sell goods illegally in Europe, you must smuggle when you come back into the colonies.

Go to the British Trade Bank to sell your goods and receive bounties. Go to the European Trade Bank if you wish to sell your goods in Europe instead of in Britain.

Step 3: Buying Goods Overseas. You can buy goods in Britain or Europe or in both to take back to America. Goods from overseas are listed on your Cargo List. If you buy illegally, you must smuggle into the colonies. Mark what you buy on your Cargo List. Put a "B" or an "E" after the goods to show where you bought them. Pay the banker. Try to choose products that give a good profit when you sell them in America.

Step 4: Returning Home. Before selling in America, you might have to pay a tax on your goods. Check the Customs board. If there are no taxes on your goods, you go through customs free and sell your goods to the American Import Bank. If there is a tax, you pay the British government.

If you do not want to pay taxes, or if you have sold or bought goods illegally, you must smuggle. Look at the American Coastline board, choose your destination, and mark your Destination Chart.

You must return to a port or smugglers' cove in your home region.

The British government can send warships or customs officials to try to catch you. Merchants and government members check their charts in secret. When you are ready, show your chart to the government. If a warship or an official is at your cove, you have been caught.

If caught by a warship, you must pay the government £300 per merchant ship. If caught by an official you might still go free. Choose a Mob or Bribe chance card. If you choose a Mob card and are lucky, a sympathetic mob scares the official away and you can sell your cargo immediately. If the mob does not support you, you go to court.

If you take a Bribe card, a corrupt official will probably let you go, but only with a large bribe. Pay bribes to the American Import Bank. An honest official will turn you over to the court.

Even if you must take a Court chance card, you might still go free or pay only a small fine. But if you are tried by a British judge, he will be hard on you. Good luck!

Step 5: Selling Goods in America. After entering the country, sell your goods from overseas to the American Import Bank. (If you were fined while short of cash, sell your cargo and then pay the fine, but the fine must be paid to the government before you go on to Step 6.)

Step 6: Buying Property and Ships. You can now invest your profits if you want. You can buy property only in your own region. Decide what you would like to own on the Investment board. Prices are on the board. Investments in property pay £100 per round profit for each round except the round in which you buy the property. You can also buy another ship by paying £400. With each new ship, you can buy three more loads of cargo.

Pay the investment banker for investments and ships. He will mark your Land Ownership Chart and put a chip on the Investment board, or if you buy a ship, he will mark your Merchant Fleet.

Properties on the Investment board are arranged in rows. You must start buying on the eastern, or right, side of the board and you cannot buy property more than one row farther west than property already owned in your region.

If you are the first to buy land in any row after Row 1, draw an Investment chance card after buying your property. The card might say your settlement has started an Indian war. It costs the British £500 to stop an Indian war. When a war occurs, the government can refuse to allow investment beyond the First Restriction Line on the board. This restriction applies for the rest of the game.

Fur buying stops for the round when a war occurs. But anyone who bought furs before the war can still sell them overseas.

At the end of each round the investment banker will pay you £100 for each investment made in previous rounds. An investment earns no profit for the first round.

RULES FOR THE BRITISH GOVERNMENT TEAM

Goals. You decide how to best govern your American colonies. You try to collect enough money to pay the expenses of governing America. To win, you must have more in your treasury at the end of the game than you started with.

Action. You start with £30,000. In each round you have a chance to collect money from the Northern, Middle, and Southern merchants. The Action Schedule shows you the steps you take in each round.

Making Money from Trade. (*Taxes:*) At the start of each round you mark on the Customs board the amount of tax on goods from overseas. Most of these are British goods. The taxes are given for the first round. One member of your team collects taxes from merchants returning to the colonies. However, some merchants may try to smuggle.

(*British transactions:*) You receive a bonus (£5 for regular loads, £30 for bounty goods) for every product bought or sold in Britain. The British trade banker will give you the bonus total at the end of the game.

(*Bounty goods:*) You especially want pitch, indigo, and timber sold in Britain. So you pay a £20 bounty on each load of these products. Give £2000 to the British trade banker to pay bounties with.

To try to keep colonial trade in Britain, the government has declared it illegal to buy or sell certain goods in certain ports. The Cargo Lists show where trade is illegal. However, many merchants want the large profits possible in illegal trade. If a merchant sells or buys illegally, he must smuggle when he returns to America.

Catching Smugglers. (*Customs officials and warships:*) You want to catch smugglers and punish them. To do this, you send warships or customs officials to the American Coastline. On your Strategy Sheet, write "O" for customs official and "W" for warship at the coves where you are sending them. Merchants check the coves they are smuggling into on their Destination Charts. Each must return to his own region.

At the start of each round you pay for warships and officials. At the most, you are allowed four warships and eight officials per region. Each warship costs £300 and each official £100. Pay the British trade banker.

When a merchant comes to the coastline, he shows you his Destination Chart. You have caught him if there is a warship or an official at his cove of entry.

If caught by a warship, the smuggler pays you £300 per merchant ship. If caught by an official, he draws a Mob or Bribe chance card. A lucky smuggler will go free or have to pay only a small bribe to the American import banker. But some will have to go to court.

(*Judges:*) The original Court cards reflect decisions of colonial juries. Often the juries free the smuggler or make him pay you only a small fine. You can increase a merchant's chances of being heavily punished by adding Judge cards to the Court chance cards. British judges fine smugglers heavily and thus bring you more money. You must pay the British trade banker £50 each round for every Judge card you use during the round.

Controlling the Indians. Whenever there is an Indian war, you must stop it. Pay the British Trade Bank £500 for troops.

When a war occurs, you can impose a restriction on further investments. Then merchants cannot buy land beyond the First Restriction Line marked on the board. Your chances of having to pay for troops will be decreased. If you want to impose the restriction, tell the investment banker not to permit any investments beyond the First Restriction Line. If investments have already gone beyond the First Restriction Line, you can restrict investments beyond the last row that has chips.

RULES FOR BANKERS

Goals and Action. You make sure that all buying and selling is done correctly. At all but one of the game's steps, merchants deal with a banker. For the game to run smoothly, you must know your job. You must also understand the action of the entire game. Read carefully the merchants' and government's rules.

The information you need for your job is printed on the board at your station.

SCORING

Merchants: The merchant with the most money wins.

British government: If you have more money in your treasury at the end of the game than you started with, you win.